TRANSITION METAL INTERMEDIATES IN ORGANIC SYNTHESIS

Transition Metal Intermediates in Organic Synthesis

by C. W. BIRD

LOGOS PRESS
ACADEMIC PRESS

© Copyright C. W. Bird 1967

Published by
LOGOS PRESS
in association with
ELEK BOOKS LIMITED

2 All Saints Street, London, N. 1.

Distributed by
ACADEMIC PRESS INC.
111 Fifth Avenue, New York, N. Y. 10003

and

ACADEMIC PRESS INC. (London) LIMITED
Berkeley Square House, Berkeley Square, London, W.1

Library of Congress Catalog Card Number 67 - 21471

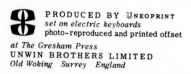
PRODUCED BY ∪NEOPRINT
set on electric keyboards
photo-reproduced and printed offset
at The Gresham Press
UNWIN BROTHERS LIMITED
Old Woking   Surrey   England

(U1863)

PREFACE

The resurgence of interest in organometallic chemistry has resulted in the discovery of many new synthetic reactions and a clearer understanding of others of longer standing. However, few of these have found their way into textbooks except in relation to specific industrial processes, and then often in small type. Thus it is hardly surprising that many organic chemists regard organometallic reagents as synonymous with Grignard reagents. It is hoped that this book will have the effect of widening their horizons in synthetic chemistry.

In deciding the scope of this book so as to avoid unnecessary duplication with other works, the field has been restricted to those reactions which involve the intermediary formation of a carbon to metal bond. Certain reactions specific to copper such as the Ullmann, Meerwein and Sandmeyer reactions have been excluded as they are well documented elsewhere. Finally, reactions of an essentially heterogeneous nature have only been discussed where they are closely related mechanistically or in synthetic application to homogeneous ones under consideration. As far as is possible I have tried to present these reactions in regard to their currently accepted mechanisms.

The discussion of carbonylation reactions in Chapters 5 to 8 is based on an earlier article published in Chemical Reviews. I am indebted to the American Chemical Society for permission to reproduce parts of this article.

C. W. BIRD

Queen Elizabeth College

University of London

November 1966

CONTENTS

CHAPTER 1

OLIGOMERISATION OF ACETYLENES

The thermal cyclisation of acetylene has long been known[12] to produce small amounts of benzene along with a wide variety of other products[9]. The discovery that acetylenes can be cyclically polymerised using transition metal catalysts is of comparatively recent origin[79].

The most extensively studied process concerns the cyclic polymerisation of acetylene in the presence of a nickel (II) compound. The principal products are cyclooctatetraene, benzene, small amounts of styrene, phenylbutadiene[22, 81], vinylcyclooctatetraene[22, 25, 39, 80], and traces of azulene and naphthalene[22, 80]. The most commonly used catalysts are derivatives of nickel (II) such as the cyanide, acetylacetonate, and ethyl acetoacetate, cf. Table 1.1. Temperatures around 100°C are normally used. An anhydrous solvent such as benzene, tetrahydrofuran or dioxan is employed as water has an adverse effect on the reaction. Adventitious moisture is removed by adding calcium carbide or some other compound which reacts rapidly with water such as ethylene oxide or an aluminium alkoxide.

In view of its previous inaccessibility the good yields of cyclooctatetraene obtained by this method encouraged the search for larger cyclic polyenes among the higher molecular weight products[79, 80]. Of particular interest was cyclodecapentaene (1) needed to explore the validity of the Hückel aromaticity rule. However, an initial misidentification of vinylcyclooctatetraene as cyclodecapentaene was soon corrected[22, 25, 39] and further search has failed to detect the desired compound. It seems possible that the elusive cyclodecapentaene is the precursor of the naphthalene and azulene. Styrene and vinylcyclooctatetraene probably originate from incorporation of

(1)

vinylacetylene which is formed in small quantities during the cyclisation reaction. Similar incorporation of the acetylene trimer hexa-1,3-dien-5-yne is probably responsible for the formation of phenyl-

1

TABLE 1.1 Cyclic Oligomerisation of Acetylenes

Acetylene	Catalyst	Products	References
Acetylene	$Ni(CN)_2$	Cyclooctatetraene, benzene, styrene	11, 16, 34, 38, 40, 77, 79, 81, 99
	$Ni_3[Cr(CN)_6]_2$	Cyclooctatetraene	33
	$Ni(NCS)_2$	Cyclooctatetraene	77
	Ni formate	Cyclooctatetraene (70%)	83
	Ni acetate	Cyclooctatetraene	54
	Ni acetylacetonate	Cyclooctatetraene (72%)	7, 8, 21, 38, 54, 74, 77, 78, 81, 99.
	Ni ethyl acetoacetate	Cyclooctatetraene	37, 38, 53, 71, 77, 81, 104, 106
	Ni *iso*-propyl acetoacetate	Cyclooctatetraene	70
	Ni *t*-butyl acetoacetate	Cyclooctatetraene	70
	Ni bornyl acetoacetate	Cyclooctatetraene	70
	Ni disalicylaldehydate	Cyclooctatetraene	38
	Ni bisacrylonitrile	Cyclooctatetraene, benzene, styrene	86, 87
	$(Ph_3P)_2Ni(CO)_2$	Benzene (63.5%), styrene (17.5%)	93
	$(Ph_4P_2)[Ni(CO)_3]_2$	Benzene	89
	$(Ph_4P_2)[Co(CO)_3]_2$	Benzene	89
	$(Ph_4P_2)[Fe(CO)_4]_2$	Benzene	89
	$(Ph_3P)Ni$ (acrylonitrile)$_2$	Benzene, styrene	87
	$(Ph_3P)_2Ni$ (acrylonitrile)$_2$	Benzene, styrene	87
	$TiCl_4 + Et_3Al$	Benzene (49%)	60

Substrate	Catalyst	Products	Ref.
Propyne	$(Ph_3P)_2Ni(CO)_2$	1, 2, 4-Trimethyl-(18%) and 1, 3, 5-trimethyl-benzenes (12%)	64
	$Et_3Al + TiCl_4$	1, 3, 5-Trimethyl-(40%) and 1, 2, 4-trimethyl-benzenes (21%)	60, 90
	$(Ph_3P)_2NiCl_2 + NaBH_4$	1, 3, 5- and 1, 2, 4-Trimethylbenzenes (15%)	29
	Vanadocene	1, 3, 5- and 1, 2, 4-Trimethylbenzenes (1 : 1.4, 33%)	95
But-1-yne	$(Ph_3P)_2Ni(CO)_2$	1, 2, 4- and 1, 3, 5-Triethylbenzenes	64
	$Et_3Al + TiCl_4$	1, 3, 5-Triethyl (35%) and 1, 2, 4-Triethyl-benzenes (17%)	60
But-2-yne	Ni bisacrylonitrile	Hexamethylbenzene	87
	$TiCl_4 + Et_3Al$	Hexamethylbenzene (80%)	60
	$TiCl_4 + i\text{-}Bu_3Al$	Hexamethylbenzene (100%)	32
Vinylacetylene	$(Ph_3P)_2Ni(CO)_2$	1, 2, 4-Trivinylbenzene	64
	$TiCl_4 + i\text{-}Bu_3Al$	1, 2, 4-Trivinyl- (67%) and 1, 3, 5-trivinyl-benzenes (7%)	45
Pent-1-yne	$(Ph_3P)_2Ni(CO)_2$	1, 2, 4- and 1, 3, 5-Tripropylbenzenes	64
	$[Co(CO)_4]_2Hg$	1, 2, 4-Tripropylbenzene (11%)	47
	Vanadocene	1, 2, 4- and 1, 3, 5-Tripropylbenzenes (1.8 : 1, 73%)	95
Isopropenylacetylene	$Ni(acac)_2 + Ph_3P$	1, 4-Dimethyl-6-isopropenylnaphthalene	18
	$(Ph_3P)_2Ni(CO)_2$	1, 4-Dimethyl-6-isopropenylnaphthalene	18
	$(Ph_3P)_2Ni(CO)_2$	1, 2, 4-Tris-(isopropenyl)benzene (10%)	64
	$(Ph_3P)Ni(CO)_3$ or $(Ph_3P)_2Ni(CO)_2$	1, 3, 5- and a little 1, 2, 4-Tris(isopropenyl)benzene	61
	$(Ph_3P)_2NiCl_2 + NaBH_4$	1, 2, 4- and 1, 3, 5-Tris(isopropenyl)benzenes (9:1) 31%	29
	$TiCl_4 + Et_3Al$	1, 3, 5-Tris(isopropenyl)benzene (87%) and a little 1, 2, 4-isomer	18, 28

(Continued overleaf)

3

TABLE 1.1 (continued)

Acetylene	Catalyst	Products	References
Hex-1-yne	$(Ph_3P)_2Ni(CO)_2$	Aromatic trimers	64
	$TiCl_4 + Et_3Al$	1,3,5-Tributylbenzene (60%)	60
	$TiCl_4 + i\text{-}Bu_3Al$	1,3,5-Tributylbenzene (92%)	32
	Vanadocene	1,2,4- and 1,3,5-Tributylbenzenes (1.9:1, 66%)	95
Hex-3-yne	$[Co(CO)_4]_2Hg$	Hexaethylbenzene (75%)	47
	$Co_4(CO)_{12}$	Hexaethylbenzene	52
	$TiCl_4 + Et_3Al$	Hexaethylbenzene (76%)	60
	$TiCl_4 + i\text{-}Bu_3Al$	Hexaethylbenzene (100%)	32
	$TiCl_4 + Al + AlCl_3$	Hexaethylbenzene	69
Hexa-1,5-diyne	$TiCl_4 + Et_3Al$	3,4,3',4',-Bisethylenebibenzyl	49
Hexa-2,4-diyne	$[Co(CO)_4]_2Hg$	1,3,5- and 1,2,4-Tri-(prop-1'-ynyl)benzene (8% and 4%)	48
	$TiCl_4 + Et_3Al$	1,3,5 Trimethyl-2,4,6-tri(prop-1'-ynyl)benzene	49
Hept-1-yne	$(Ph_3P)_2Ni(CO)_2$	Aromatic trimers	64,66
Hepta-1,6-diyne	$(Ph_3P)_2Ni(CO)_2$	1,3-Di-(5-indanyl)propane	1,19
	$TiCl_4 + Et_3Al$	1,3-Di-(5-indanyl)propane	49
Oct-1-yne	$TiCl_4 + Al + AlCl_3$	1,3,5- and 1,2,4-Trihexylbenzene (56%)	69
Di-isopropenylacetylene	$Co_2(CO)_8$	Hexakis-(isopropenyl)benzene (4%)	4
Octa-1,7-diyne	$[Co(CO)_4]_2Hg$	6-(hex-5'-ynyl)tetrahydronaphthalene (2%)	49
	$TiCl_4 + Et_3Al$	1,4-Di-(6-tetrahydronaphthyl)butane (13%)	49
Nona-1,8-diyne	$[Co(CO)_4]_2Hg$	1-(Hept-6'-ynyl)-3,4-pentamethylenbenzene	49

Substrate	Catalyst	Product	Reference
	$TiCl_4 + Et_3Al$	1,5-Di-(3,4-pentamethylenebenzene)pentane (12%) and (24, n = 5) (1%)	49
Dec-5-yne	$TiCl_4 + Et_3Al$	Hexabutylbenzene (52%)	60
Deca-1 9-diyne	$TiCl_4 + Et_3Al$	(24, n = 6) (0.2%)	49
Undeca-1,10-diyne	$TiCl_4 + Et_3Al$	(24, n = 7) (0.3%)	49
Undeca-2,9-diyne	$[Co(CO)_4]_2Hg$	1,2,4-Trimethyl-3-oct-6'-ynyl-5,6-pentamethylenebenzene	49
Phenylacetylene	$(Ph_3P)_2Ni(CO)_2$	1,2,4- and 1,3,5-Triphenylbenzene (20 and 1%)	64
	$[o\text{-}(Et_2P)_2C_6H_4]Ni(CO)_2$	1,3,5-Triphenylbenzene	17
	$[Co(CO)_4]_2Hg$	1,2,4-Triphenylbenzene (70%)	47
	$Co_2(CO)_6(C_6H_5 \cdot C_2 \cdot H)$	1,2,4-Triphenylbenzene	52
	$TiCl_4 + Et_3Al$ or $i\text{-}Bu_3Al$	1,3,5-Triphenylbenzene (90%)	35, 42, 76
	$TiCl_4 + Al + AlCl_3$	1,2,4- and 1,3,5-Triphenylbenzenes	69
	$(n\text{-}Bu_3P)_2NiCl_2 + NaBH_4$	1,2,4- and 1,3,5-Triphenylbenzenes	58
	$(Ph_3P)_2NiCl_2 + NaBH_4$	1,2,4- and 1,3,5-Triphenylbenzenes (3.3:1, 50%)	29
p-Bromophenylacetylene	$[Co(CO)_4]_2Hg$	1,2,4-Tris-(p-bromophenyl)benzene (65%)	47
p-Nitrophenylacetylene	$(Ph_3P)_2Ni(CO)_2$	Tris-(p-nitrophenyl)benzene (19%)	64
1-Phenylprop-1-yne	$[Co(CO)_4]_2Hg$	1,2,4-Trimethyl-3,5,6-triphenylbenzene (90%)	47, 52
1-Phenylpenta-1,3-diyn-5-ol	$(Ph_3P)Ni(CO)_3$	1,2,4-Tris-(Phenylethynyl)-3,5,6-tris-(hydroxymethyl)benzene (40%)	48
1-Phenyl-5-methoxypenta-1,3-diyne	$(Ph_3P)Ni(CO)_3$	1,2,4-Tris-(Phenylethynyl)-3,5,6-tris-(methoxymethyl)benzene (2%), 1-phenyl-3,5-bis(methoxymethyl)-2,4-bis(phenylethynyl)-6-(3'-methoxypropynyl)benzene (7%)	48
	$[Co(CO)_4]_2Hg$	1,3,5-Tris-(phenylethynyl)-3,5,6-tris (hydroxymethyl)benzene (3%)	48

(Continued overleaf)

TABLE 1.1 (continued)

Acetylene	Catalyst	Products	References
1-Chlorophenylacetylene	$[Co(CO)_4]_2Hg$	1, 2, 4-Trichloro-3, 5, 6-triphenylbenzene (14%)	47
Diphenylacetylene	Ni bisacrylonitrile	Hexaphenylbenzene (48%), 2, 3, 4, 5-tetraphenylbenzonitrile (29%)	87
	Ni bisacraldehyde	Hexaphenylbenzene (62%), 2, 3, 4, 5-tetraphenylbenzaldehyde	87
	$[Co(CO)_4]_2Hg$	Hexaphenylbenzene (90%)	47, 52
	$Co_4(CO)_{12}$	Hexaphenylbenzene (66%)	52
	$Co_2(CO)_6(Ph_2C_2)$	Hexaphenylbenzene (84%)	52
	cycloheptatriene $Cr(CO)_3$	Hexaphenylbenzene (59%)	92
	benzene $Cr(CO)_3$	Hexaphenylbenzene (43%)	92
	toluene $Cr(CO)_3$	Hexaphenylbenzene (25%)	92
	hexamethylbenzene $Cr(CO)_3$	Hexaphenylbenzene (27%)	92
	naphthalene $Cr(CO)_3$	Hexaphenylbenzene (22%)	92
	p-dimethoxybenzene $Cr(CO)_3$	Hexaphenylbenzene (10%)	92
	chlorobenzene $Cr(CO)_3$	Hexaphenylbenzene (20%)	92
	p-dichlorobenzene $Cr(CO)_3$	Hexaphenylbenzene (11%)	92
	cycloheptatriene $Mo(CO)_3$	Hexaphenylbenzene (32%)	92
	$(PhCN)_2PdCl_2$	Hexaphenylbenzene (90%)	13, 62
	$TiCl_4 + Et_3Al$	Hexaphenylbenzene	30
	$TiCl_4 + i\text{-}Bu_3Al$	Hexaphenylbenzene (100%)	32
	$TiCl_4 + LiAl(C_7H_{15})_4$	Hexaphenylbenzene	2
	$TiCl_4 + Al + AlCl_3$	Hexaphenylbenzene	69
Di-p-chlorophenylacetylene	$[Co(CO)_4]_2Hg$	Hexakis-(p-chlorophenyl)benzene (95%)	47

Substrate	Catalyst	Products	Ref.
Diphenylbutadiyne	$[Co(CO)_4]_2Hg$	1,3,5- and 1,2,4-Tris-(phenylethynyl)triphenyl-benzenes (21 and 13%)	48
Ethoxyacetylene	$(Ph_3P)_2Ni(CO)_2$	1,2,4-Triethoxybenzene (29%)	64
Propargyl alcohol	$(Ph_3P)_2Ni(CO)_2$	1,2,4- and 1,3,5-Trimethylolbenzenes (42 and 23%)	64, 82
	$[(NC.C_2H_4)_3P]_2NiCl_2 + NaBH_4$	1,3,5- and 1,2,4-Trimethylolbenzenes (65 and 35%)	58
1-Methoxyprop-2-yne	$Co_2(CO)_6(methoxyprop-2-yne)$	1,2,4-Tris-(methoxymethyl)benzene (17%)	47
1-Phenoxyprop-2-yne	$(Ph_3P)_2Ni(CO)_2$	1,2,4- and 1,3,5-Tris-(phenoxymethyl)benzenes (63%)	64
But-3-ynol	$(Ph_3P)_2Ni(CO)_2$	Tris-(β-hydroxyethyl) benzenes (38%)	64
	$Co_2(CO)_6(but-3-ynol)$	1,2,4-Tris-(β-hydroxyethyl) benzene (14%)	47
But-2-yn-1,4-diol	$(Ph_3P)Ni(CO)_3$	Hexamethylolbenzene (55%)	50
2-Methylbut-3-yn-2-ol	$(Ph_3P)Ni(CO)_3$ or $(Ph_3P)_2Ni(CO)_2$	1,3,5-Tris-(α-hydroxyisopropyl)benzene (61%)	61
	$(Ph_3P)Ni(CO)_3$	1,2,4-Tris-(α-hydroxyisopropyl)benzene (63%)	91
	$Co(CO)_3NO$	1,2,4-Tris-(α-hydroxyisopropyl)benzene (95%)	91
1-Methoxybut-1-en-3-yne	$(Ph_3P)_2Ni(CO)_2$	1,3,5-Tris-(β-methoxyvinyl)benzene	64
Phenylprop-2-yn-1-ol	$(Ph_3P)_2Ni(CO)_2$	1,2,4-Tris-(α-hydroxybenzyl)benzene	84
Butyn-3-one	$[(NC.C_2H_4)_3P]_2Ni(CO)_2$	1,3,5-Triacetylbenzene (85%)	64
Methyl propiolate	$(Ph_3P)_2Ni(CO)_2$	1,2,4-Tricarbomethoxybenzene (78%)	64
	$(PCl_3)_4Ni$	1,2,4,6-Tetracarbomethoxycyclooctatetraene (83%), 1,2,4-tricarbomethoxybenzene (17%)	55, 56
Ethyl propiolate	$(Ph_3P)_2Ni(CO)_2$	1,2,4- and 1,3,5-Tricarbethoxybenzenes (89 and 6%)	64, 66
	$(PCl_3)_4Ni$	1,2,4,6- and 1,3,5,7-Tetracarbethoxycyclooctatetraenes (28 and 1%), 1,2,4- and 1,3,5-tricarbethoxybenzenes (71%)	55, 56

7

TABLE 1.1 (continued)

Acetylene	Catalyst	Products	References
Phenylpropiolic acid	$Co_2(CO)_6(Ph.C_2.CO_2H)$	1,2,4-Triphenylbenzene-3,5-6-tricarboxylic acid (11%)	47
Methyl phenylpropiolate	$[Co(CO)_4]_2Hg$	1,2,4-Tricarbomethoxy-3,5,6-triphenyl-benzene (55%)	47,52
	$Co_4(CO)_{12}$	1,2,4-Tricarbomethoxy-3,5,6-triphenyl-benzene	52
Dimethyl acetylenedicarboxy-late	$[Co(CO)_4]_2Hg$	Methyl mellitate (80%)	47,52
1-Dialkylaminoprop-2-ynes	$(Ph_3P)_2Ni(CO)_2$	1,3,5-Tris-(dialkylaminomethyl)benzenes	100
Trimethylsilylacetylene	$[Co(CO)_4]_2Hg$	1,2,4-Tris-(trimethylsilyl)benzene	52
	$Co_2(CO)_6(Me_3SiC_2H)$	1,2,4-Tris-(trimethylsilyl)benzene (55%)	52
	$TiCl_4 + Me_3Al$	1,3,5- and 1,2,4-Tris-(trimethylsilyl) benzenes (80 and 20%)	26
1-Phenyl-4-trimethylsilylybuta-diyne	$[Co(CO)_4]_2Hg$	1,3,5-Tris-(trimethylsilylethynyl)-2,4,6-triphenylbenzene (50%)	48
Bis-(trimethylsilyl)butadiyne	$[Co(CO)_4]_2Hg$	1,3,5-Tris-(trimethylsilylethynyl)-2,4,6-tris (trimethylsilyl) benzene (20%), 1,2,4-tris-(trimethylsilylethynyl)-3,5,6-tris-(trimethylsilyl) benzene (25%)	48

butadiene. A number of other acetylenes have been copolymerised with acetylene producing substituted cyclooctatetraenes but usually only one substituted acetylene unit has been incorporated[20, 24], Table 1.2. The nickel (II) catalyst system is apparently unable to oligomerise substituted acetylenes.

Longuett-Higgins and Orgel[57] suggested that a cyclobutadiene nickel (II) complex was an intermediate in the cyclic polymerisation of acetylene. Dimerisation of cyclobutadiene to cyclooctatetraene could explain the preponderance of cyclooctatetraene over benzene. The generation of cyclobutadiene by oxidation of its iron tricarbonyl complex[31] with ceric sulphate has enabled a study of its reactions to be made[10, 15, 101, 102]. Some of these are summarised in Figure 1. In particular it may be noted that dimerisation only occurs in the absence of an alternative acetylene or olefin. Thus if cyclobutadiene were an intermediate in catalytic cyclooctatetraene formation benzene would be expected as the major product, in view of the large excess of acetylene present. For this reason, at least, it seems unlikely that cyclobutadiene is involved in this particular cyclisation reaction.

The relationship between the structure and catalytic activity of nickel (II) complexes for cyclooctatetraene formation has been investigated[38, 88]. All of the active catalysts are characterised by having weak or intermediate range ligand fields and show paramagnetic or anomalous magnetic behaviour. The most important property of these octahedral or planar complexes in the present context is their ability to undergo rapid ligand exchange. In contrast planar complexes such as nickel dimethylglyoxime or phthalocyanine with strong ligand fields and very slow rates of ligand exchange are catalytically inactive. The importance of ease of ligand exchange is further emphasised by the bis-(NN'-dialkylaminotroponiminato) nickel (II) complexes (2). Complex (2a) is diamagnetic and planar

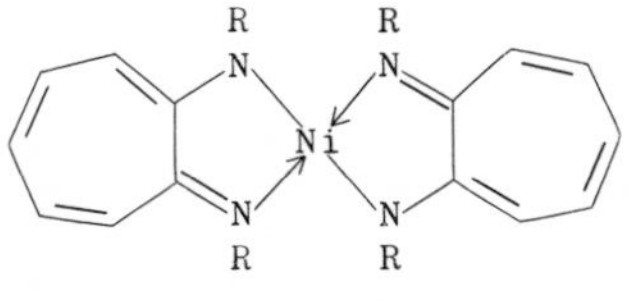

2 (a) R = CH_3

(b) R = C_2H_5

and (2b) paramagnetic and tetrahedral. Both complexes undergo ligand exchange very slowly in inert solvents and show negligible activity for cyclisation of acetylene.

The influence of these factors has been further illustrated with the bis(N-alkylsalicylaldiminato)nickel (II) complexes (3). The magnetic

10

TABLE 1.2 Cyclic Co-oligomerisation of Acetylenes

Acetylenes	Catalyst	Products	References
Acetylene + Propyne	Ni acetylacetonate	Monomethyl (16%) and dimethylcyclo-octatetraenes, toluene and benzene	20
	$(Ph_3P)_2NiCl_2 + NaBH_4$	m-Xylene and toluene (14%, 1:4)	29
Acetylene + But-2-yne	Ni acetylacetonate	1, 2-Dimethylcyclooctatetraene (19%), cyclooctatetraene, o-xylene, benzene	20
	$(Ph_3P)_2Ni(CO)_2$	o-Xylene (50%), benzene (43%), styrene (3.5%)	64, 85
Acetylene + Vinylacetylene	Ni ethyl acetoacetate	Vinylcyclooctatetraene	22, 25, 38
	$(Ph_3P)_2Ni(CO)_2$	Benzene (50-60%), styrene (30-20%)	105
	$Ph_3P[Ni(CO)_2]_3$?	Divinylbenzene and styrene	72
Acetylene + Pent-1-yne	Ni acetylacetonate	n-Propylcyclooctatetraene (25%), cyclo-octatetraene, n-propylbenzene, benzene	20
Acetylene + Hex-1-yne	Ni acetylacetonate	n-Butylcyclooctatetraene (16%), cyclo-octatetraene, benzene	20
Acetylene + Divinylacetylene	$(Ph_3P)Ni(CO)_3$	1, 2-Divinylbenzene (36%)	23, 45
	$TiCl_4 + i$-Bu_3Al	1, 2-Divinylbenzene (30%)	45
Acetylene + Di-t-butylacetylenedicobalt hexacarbonyl		1, 2-Di-t-butylbenzene	5, 44
Acetylene + Phenylacetylene	Ni acetylacetonate	Phenylcyclooctatetraene (17%)	20
	$[Co(CO)_4]_2Hg$	p-Terphenyl (3%), 1, 2, 4-triphenylbenzene (3%)	47
Acetylene + Diphenylacetylene	Ni acetylacetonate	1, 2-Diphenylcyclooctatetraene (14%), cycloocta-tetraene, 1, 2-diphenylbenzene (6%), benzene	24
Propyne + Divinylacetylene	$TiCl_4 + i$-Bu_3Al	Dimethyl-o-divinylbenzenes (28%), trimethyl-benzenes	45
Pent-1-yne + Hepta-1, 6-diyne	$(Ph_3P)_2Ni(CO)_2$	5-Propylindane	19
Hex-1-yne + Phenylacetylene	$TiCl_4 + i$-Bu_3Al	Tributylbenzene, dibutylphenylbenzene, butydiphenylbenzene, triphenylbenzene (1:3:3:1)	75

Hex-3-yne + Diphenylacetylene	$[Co(CO)_4]_2Hg$	Hexaphenylbenzene (15%), 1,2-diethyltetra-phenylbenzene (57%), 1,2-diphenyltetraethylben-zene (7%)	47
t-Butylacetylene + Di-t-butylacetylenedicobalt hexacarbonyl		1,2,4,5-Tetra-t-butylbenzene	6
t-Butylacetylene + Diphenylacetylenedicobalt hexacarbonyl		Di-t-butyldiphenylbenzene (36%)	52
Hept-1-yne + Hepta-1,6-diyne	$(Ph_3P)_2Ni(CO)_2$	5-Pentylindane	19
Hept-1-yne + Octa-1.7-diyne	$(Ph_3P)_2Ni(CO)_2$	6-Pentyltetralin	19
Di-t-butylacetylene + t-Butylacetylenedicobalt hexacarbonyl		1,2,4-Tri- and 1,2,4,5-tetra-t-butylbenzenes	43
Phenylacetylene + Diphenylacetylene	$[Co(CO)_4]_2Hg$	Hexaphenylbenzene (34%), pentaphenylbenzene (1%), 1,2,3,5-tetraphenylbenzene (12%)	47
Diphenylacetylene + Di-(p-chlorophenyl)acetylene	$[Co(CO)_4]_2Hg$	Hexaphenyl-, 1,2-di-(p-chlorophenyl)tetra-phenyl-, 1,2-diphenyltetra-(p-chlorophenyl)-and hexa(p-chlorophenyl)benzenes (1:3:3:1)	47
Diphenylacetylene + Dimethyl acetylene-dicarboxylate	$[Co(CO)_4]_2Hg$	Methyl mellitate (16%), 1,2-diphenyl-3,4,5,6-,tetracarbomethoxybenzene (10%), dimethyl tetraphenylphthalate (0.1%)	47
Trimethylsilylacetylene + Diphenylacetylenedicobalt hexacarbonyl		Bis-(trimethylsilyl)diphenylbenzene (7%) 1,2,4-trimethylsilylbenzene (47%)	52
Bis-(trimethylsilyl) acetylene + Diphenylacetylene-dicobalt hexacarbonyl		1,2-Bis-(trimethylsilyl)tetraphenylbenzene	52
Phenyltrimethylsilylacetylene + Diphenylacetylenedicobalt hexacarbonyl		1,3-Bis-(trimethylsilyl)tetraphenylbenzene, 1,3,5-triphenyl-2,4,6-tris (trimethylsilyl)benzene	52

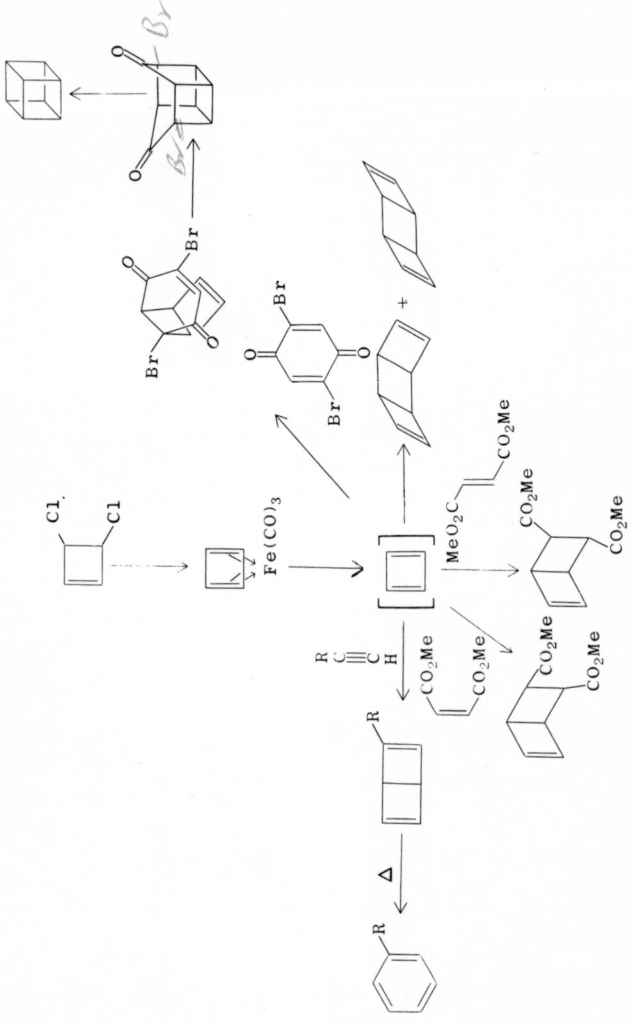
Cl
Cl
Fe(CO)3
Br
Br
Br
Br
MeO2C
CO2Me
CO2Me
CO2Me
CO2Me
CO2Me
CO2Me
CO2Me
R
C
C
H
R
R
Δ

FIGURE 1.1

moments and catalytic activities, for both cyclooctatetraene and benzene formation, of a series of these complexes with $R = n$-alkyl
closely parallel one another, with the most active having $R =$ methyl.
Equally, the catalytic activity of these complexes parallels their
relative rates of exchange with dimethylglyoxime in the same solvent (Figure 1. 2). These replacement reactions with dimethylglyoxime are second order, the rate being controlled by the addition of
the first molecule of dimethylglyoxime to the nickel chelate. Steric
effects are also observed in chelates (3) with branched alkyl groups.

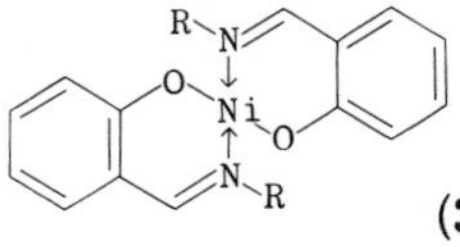

(3)

Again the yields of cyclooctatetraene are roughly proportional to the
rate of ligand exchange with dimethylglyoxime, but the yield of
benzene is proportionately higher than when using (3) with straight
chain alkyl groups. Ligand mobility is also an important factor with
these chelates as complete inactivity results in complexes of type
(4) when $n = 6$, but slight activity is observed for $n = 7$ or 8.

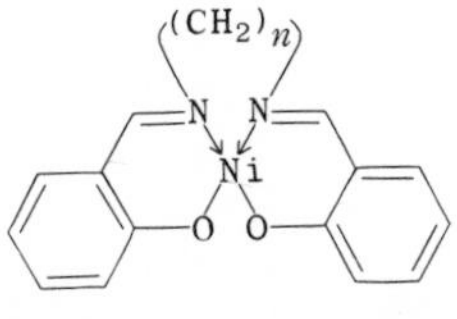

(4)

The foregoing evidence suggests that cyclooctatetraene is formed
within a nickel-acetylene complex produced by replacement of the
ligands with acetylene. In the resultant octahedral complex, e.g. (5),
the acetylene molecules can adopt a conformation favourable to the
formation of cyclooctatetraene.

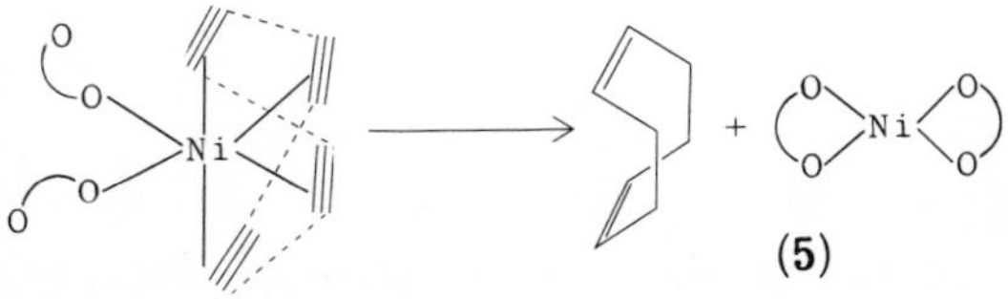

(5)

Additional circumstantial evidence is provided by the similarity of
the solvent effects for both cyclooctatetraene formation and ligand

13

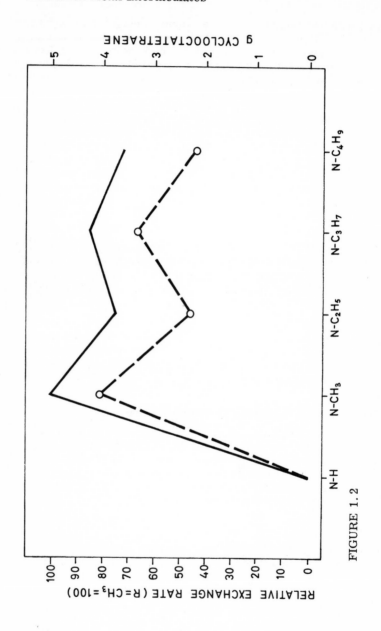

FIGURE 1.2

exchange reactions with these chelates. Both processes proceed more rapidly in polar solvents. The yields of cyclooctatetraene are larger in dioxan than in tetrahydrofuran and are least in benzene. More polar solvents such as pyridine, water or benzonitrile may inhibit the reaction by blocking the coordination of the acetylene molecules.

The mechanism envisaged for this reaction suggests that the course of cyclisation can be modified in the presence of suitable ligands.

Incorporation of triphenylphosphine to the extent of one mole per mole of catalyst blocks the formation of cyclooctatetraene in favour of benzene, which is obtained in high yield. Larger amounts of triphenylphosphine do not inhibit benzene formation, so that triphenylphosphine must be occupying only one of the four coordination positions. The introduction of bidentate ligands such as 1, 2-bis-(diphenylphosphino) ethane, α, α'-dipyridyl or 1, 10-phenanthroline, which can block two cis positions in the active complex, completely deprives it of all catalytic activity. A series of ligands can be arranged in order of increasing inhibiting effect on cyclooctatetraene formation: trimethylamine < dimethylaniline < pyridine < triphenylbismuth < triphenylstibine < triphenylarsine < triphenylphosphine < diphenylethylphosphine < tributylphosphine[38, 106]. The order of this series throws some light on the bonding of acetylene in the intermediate complex. Since even trimethylamine or pyridine, which are σ-donors and only very weak π-donors, inhibit cyclooctatetraene formation it can be concluded that only relatively weak σ-bonds with small π-bond contributions are formed between acetylene and the nickel (II) ion.

The modifying ligand may be incorporated into the chelate as in (6). Although the activity of these complexes is not very high, a higher

(6)

proportion of benzene, relative to cyclooctatetraene is formed for $n = 4$ than with $n = 2, 3, 5$ or 6. This may correspond to the presence of an optimal ring size for chelation of the oxygen with the nickel ion.

Only one other group of catalysts, which are complexes of nickel (0), have been found to convert acetylenes into cyclooctatetraenes. One of the most recently discovered examples is bisacrylonitrile nickel (0), which is a coordinatively unsaturated π-complex of remarkable catalytic activity. It very smoothly converts acetylene into benzene and cyclooctatetraene[86, 87], but gives only benzenoid derivatives with substituted acetylenes. The introduction of ligands such as triphenylphosphine increases the yield of benzene but completely suppresses cyclooctatetraene formation[87].

The sole substituted acetylenes which have been tetramerised to cyclooctatetraene derivatives are methyl and ethyl propiolate. Using tetrakis (phosphorus trihalide) nickel (0) complexes[55, 56] ethyl propiolate produces 1, 2, 4, 6-(7, R = C_2H_5) 28% and 1, 3, 5, 7-tetra-carbethoxycyclooctatetraene (8) 1% in addition to a mixture (71%) of 1, 2, 4 and 1, 3, 5-tricarbethoxybenzene. Methyl propiolate differs in giving only one tetramer (7, R = CH_3) (83%) and 1, 2, 4-tricarbomethoxybenzene (17%). The decreasing order of catalyst efficiency is $Ni(PCl_3)_4 > Ni(PF_3)_4 \gg Ni(PBr_3)_4 \sim Ni(PhPCl_2)_4$. A large range of other acetylenes were found to be unreactive.

(7)   (8)

One of the earliest examples of a nickel (0) complex to be used for the more general cyclic polymerisation of acetylenes was dicarbonyl-bis(triphenylphosphine)nickel (0)[82]. The reaction is carried out in an inert solvent such as benzene at reflux. In the case of reactive acetylenes the reaction may be strongly exothermic. In all cases the solution darkens and most of the carbon monoxide is evolved. The original nickel complex is finally converted into triphenylphosphine oxide and nickel metal, oxide or hydroxide. The products depend on the acetylene used and are substituted benzenes together with linear condensation polymers[64], (Table 1. 1). The formation of these linear polymers, which are the main products from the less reactive acetylenes, was not noticed in the earlier work and has led to much confusion, as in the case of phenylacetylene[14, 73, 84]. A recent extensive survey has helped to clarify the position[64].

(9)   (10)

Most disubstituted acetylenes and monosubstituted acetylenes with l-butyl, carboxyl, amide, nitrile or halogen substituents, are unreactive under these conditions. Benzenoid compounds are the main products from mono-substituted acetylenes with lower alkyl, aryl, vinyl, hydroxymethyl, alkoxycarbonyl, acyl and alkoxyl substituents. The 1, 2, 4-trisubstituted benzene usually predominates over the 1, 3, 5-isomer. Monosubstituted acetylenes with higher alkyl, cyclohexyl and hydroxyalkyl substituents give linear oligomers. These are principally dimers (9 and 10) or trimers (11 and 12) of the acetylene[63].

(11)   (12)

The relative rates of linear polymerisation and cyclotrimerisation remain constant throughout the reaction and both have the same hydrogen deuterium isotope effect of 2.5 for terminally deuterated acetylenes[66]. A hypothetical mechanism has been proposed to account for these and the following observations.

A detailed study[66] of the cyclic trimerisation of ethyl propiolate and the linear oligomerisation of hept-1-yne has helped to define the optimum reaction conditions and catalyst. Firstly the reactions are fairly independent of the solvent used except where it has strong coordinating properties or destroys the catalyst. The sensitivity of the reaction to temperature and agitation seems to arise from the need to expel liberated carbon monoxide, which has a strong inhibitory effect. Pretreatment of the catalyst with a disubstituted acetylene, which is not itself trimerised, enhances the reactivity of the catalyst. This activation is accompanied by liberation of carbon monoxide and formation of an intense yellow air sensitive species, which may be of the bisphosphinenickel—acetylene type[103]. The optimum acetylene to catalyst ratio is below 1,000 to 1. No relationship appears to exist between the induction period and the catalyst efficiency. In a series of complexes of the type $(R_3P)_2 Ni(CO)_2$ the activity varied with R as $NC.CH_2.CH_2 > C_6H_5 > H > OC_2H_5 > n$-alkyl $\gg O.C_6H_5 \sim Cl$. Exchange of tertiary phosphine (R_3P) with this type of complex proceeds through an S_N1 dissociative mechanism and the rate of dissociation decreases in the order $R = Cl \gg C_6H_5 \sim NC.CH_2.CH_2 > n$-butyl $\gg OC_2H_5 \sim OC_6H_5$[65]. This supports the conclusion that it is the carbon monoxide ligands which are displaced in the activation process and not the phosphines. It may be noted that nickel carbonyl is generally inactive and that $(R_3P) Ni (CO)_3$ complexes give slightly better conversions than $(R_3P)_2 Ni(CO)_2$ complexes. Product distribution is also sensitive to catalyst structure. Triphenylphosphite containing complexes favour formation of

17

benzenoid products, whereas tris(2-cyanoethyl)phosphine complexes promote linear polymerisation.

The principal disadvantages of the nickel (0) catalysts discussed above are their propensity for producing linear oligomers and their low reactivity towards disubstituted acetylenes. These drawbacks can be circumvented by the use of other metal carbonyls. A general survey has been made of the effectiveness of a variety of such compounds for the conversion of diphenylacetylene into hexaphenylbenzene, Table 1.3. Cobalt carbonyl complexes such as bis(tetracarbonylcobalt)mercury are particularly effective. Good yields of hexaphenylbenzene may be obtained in a wide variety of solvents, dioxan being one of the best[47]. It may be noted that with these catalysts also unsymmetrical acetylenes are converted predominantly into the 1, 2, 4-rather than the 1, 3, 5-trisubstituted benzene.

TABLE 1.3 Cyclic Trimerisation of Diphenylacetylene[47]

Catalyst	Temperature (°C)	% Yield of Hexaphenylbenzene
$Fe_3(CO)_{12}$	260-280	75
$Fe_2(CO)_9$	250	25
$Fe(CO)_5$	270	20
$Fe(CO)_4Hg$	285	50
$Fe_2(CO)_6(Ph_2C_2)_2$	270	60
$Fe_2(CO)_7(Ph_2C_2)_2$	250–270	15
$Fe_3(CO)_8(Ph_2C_2)_2$	260–280	50
(Tetracyclone) $Fe(CO)_3$	280	1
$[C_5H_5Fe(CO)_2]_2$	270	6
$Co_2(CO)_8$	280	60
$[Co(CO)_4]_2Hg$	270	70
$Co_2(CO)_6(Ph_2C_2)$	150	70
$Ni(CO)_4$	260	5
$Mn_2(CO)_{10}$	270	55
$Mo(CO)_6$	270	50
$W(CO)_6$	270	15

An insight into the mechanism of the cobalt carbonyl catalysed cyclisations has been provided by an extensive investigation of their reactions with acetylenes. Typically, diphenylacetylene reacts with dicobalt octacarbonyl to give the diphenylacetylene dicobalthexacarbonyl complex (13), which with further diphenylacetylene yields hexa-

(13)

phenylbenzene, the ubiquitous tetracyclone and two further complexes. The structures of these complexes are probably (14) and (15). The latter structure is assigned by analogy with that found for a similar complex[68]. Complex (15) is decomposed by air or heating to hexaphenylbenzene. The reaction of diphenyl-acetylene with tetracobalt dodecacarbonyl or bis(tetracarbonylcobalt) mercury also forms hexaphenylbenzene and complexes (14) and (15). In reactions involv-

(14)

(15)

ing unsymmetrical acetylenes such as methyl phenylpropiolate the unsymmetrical benzene, i.e. 1, 2, 4-tricarbomethoxy-3, 5, 6-triphenylbenzene is isolated directly. Degradation of the complex analogous to (15) gives the symmetrical 1, 3, 5-tricarbomethoxy-2, 4, 6-triphenylbenzene. Obviously the complex leading to the unsymmetrical benzene is much less stable, possibly for steric reasons, than the one leading to the symmetrical benzene. Degradation of such complexes either thermally or with bromine can occasionally provide a useful preparative method[51, 52].

The problem of determining the relative reactivities of acetylenes in cyclotrimerisation reactions catalysed by bis(tetracarbonylcobalt) mercury has been solved by cotrimerising pairs of acetylenes and determining the relative amounts of benzenes formed[47]. Two acetylenes, X_2C_2 and Y_2C_2, of comparable reactivity will give the four possible isomers C_6X_6, $C_6X_4Y_2$, $C_6X_2Y_4$ and C_6Y_6 in the statistical ratio of 1:3:3:1, whereas if one is less reactive then proportionately smaller amounts are formed of the benzenes derived from one, two and three molecules of the less reactive acetylene. On this basis the order of reactivity of a series of acetylenes can be deduced as dimethyl acetylenedicarboxylate > diphenylacetylene ~ 4, 4′-dichlorodiphenylacetylene > phenylacetylene > hex-3-yne > acetylene.

The cobalt carbonyl catalytic system is the only one which, up to the time of writing, has produced results explicable only if a cyclobutadiene intermediate has been formed at some stage. The reaction of diphenylacetylenedicobalt hexacarbonyl (13) with bis(trimethylsilyl) acetylene produces the following products[52].

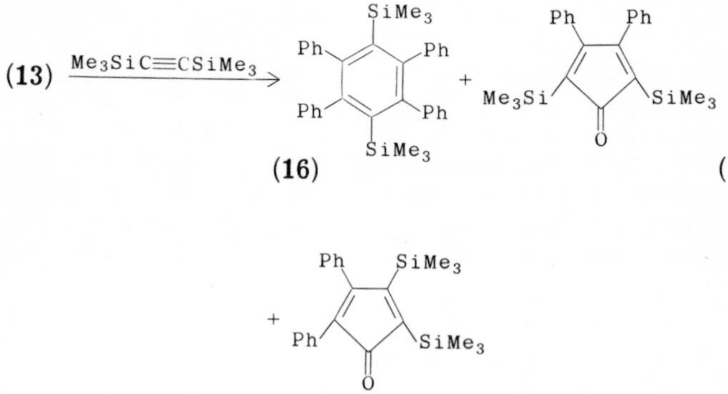

(16)   (17)

(18)

The formation of (16) and (17) must have involved an intermediate with at least the symmetry of cyclobutadiene. The reaction of diisopropylacetylene with dicobalt octacarbonyl also produces products (21) and (22) suggestive of a cyclobutadiene precursor[3].

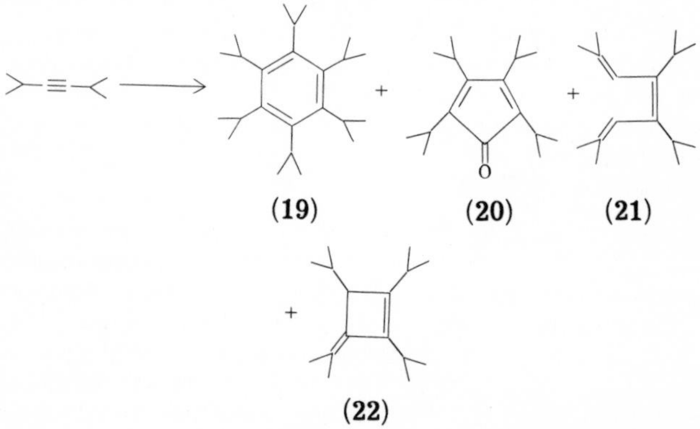

(19)   (20)   (21)

(22)

These results indicate that a little caution may be necessary in assigning orientations to benzenes obtained by this route.

The cyclic trimerisation of acetylenes to benzenes can also be effected with Ziegler-type catalysts. This approach is essentially complementary to the foregoing methods as unsymmetrical acetylenes give preponderantly the symmetrical benzene. The catalyst system comprises a trialkylaluminium and titanium tetrachloride with an aluminium to titanium ratio between 1:1 and 3:1[32, 60]. Outside these limits only polymers are formed. High dilution techniques must be used with monoalkyl acetylenes[60]. Substitution of titanium trichloride for tetrachloride also leads to polymer formation[32, 75, 76]. The presence of potential ligands such as tetrahydrofuran or pyridine hinders or stops the reaction[75].

Most of the simpler acetylenes cyclotrimerised using the Ziegler-type catalyst have given excellent yields of the corresponding benzene. α, ω-Diacetylenes, $HC = C.(CH_2)_nC = CH$, with this catalyst give products of the type (23)[49]. The yields of (23, $n = 2$ to 6) are

(23)

better when the reaction is carried out at high dilution, for example when $n = 2$ it is increased from 2 to 13%. The best yield is obtained when $n = 4$ and then falls off as n is increased. An additional product when $n = 5$ to 7 is (24), which is obtained in less than 1% yield.

(24)

Compounds of the type (23) are also obtained from α, ω-diacetylenes using bis(triphenylphosphine)nickel dicarbonyl[1, 19], but bis(tetracarbonylcobalt)mercury gives instead compounds of type (25)[49]. A similar type of catalyst is obtained by treatment of a bisphosphine

(25)

nickel (II) halide with sodium borohydride in water, alcohol or acetonitrile[29, 36, 58, 59]. If diglyme is used as a solvent lithium aluminium hydride is a satisfactory reductant[59]. The nature of the halide
is immaterial but tris-(2-cyanoethyl)phosphine is superior to tri-
phenyl- or tributylphosphine. The actual catalytic species has been
identified[36] as a metal hydride, e.g. $(R_3P)_2NiHCl$. It has also been
reported[27] that bis-phosphine nickel halides are capable of effecting
such polymerisation without the addition of a reducing agent. In this
case the activity depended on the halide with iodide $>$ bromide $>$
chloride suggesting that reduction is being effected by the reactants.

These systems convert acetylene into polyacetylene. Substituted
acetylenes are converted into a mixture of predominantly dimers,
trimers and tetramers. Both linear and aromatic products are produced, the actual proportions depending on the individual acetylene.
Monosubstituted acetylenes give a larger proportion of the 1, 2, 4-
isomer than obtained with the titanium catalyst[29]. However, the
greater practical ease with which this catalytic system can be used
in many cases more than compensates for both this and the somewhat lower yields. Similar results are obtained using cobalt (II) complexes with borohydride[36, 59], whereas palladium, platinum, osmium
and ruthenium halides with borohydride give a much larger dimer to
trimer ratio[59]. It is interesting that compounds of iron and other
transition metals showed no catalytic activity under these conditions.

Aryl and alkyl derivatives of transition metals have been found to
convert acetylenes into aromatic hydrocarbons. While the use of
systems of this type has not been so extensively studied as the catalysts discussed so far, the products are often different as a result of
the incorporation of the aryl or alkyl groups.

Triphenylchromium reacts exothermically with but-2-yne in tetrahydrofuran, after an induction period whose length is dependent on
the relative quantities of reactants[41, 107]. The products depend on
the ratio of but-2-yne to triphenylchromium. At a ratio of 4:1 or
less 1, 2, 3, 4-tetramethylnaphthalene and its chromium π-complex
are obtained. A 6:1 ratio gives 1, 2, 3, 4-tetramethylnaphthalene
(40%) and hexamethylbenzene (50%). Even higher ratios give the
same products accompanied by their chromium π-complexes. The
variations in product have been interpreted as supporting the stepwise replacement of tetrahydrofuran by but-2-yne, Figure 1. 3.

Tri-α-naphthylchromium converts but-2-yne into hexamethylbenzene and 1, 2, 3, 4-tetramethylphenanthrene. The same products and
smaller amounts of 1, 2, 3, 4-tetramethylanthracene were obtained
using tri-β-naphthylchromium. Analogously, trivinylchromium
yields hexamethylbenzene and 1, 2, 3, 4-tetramethylbenzene[94]. Tri-
mesitylchromium, where cyclisation on to the mesityl group is impossible, converts but-2-yne into hexamethylbenzene and 2-mesityl-

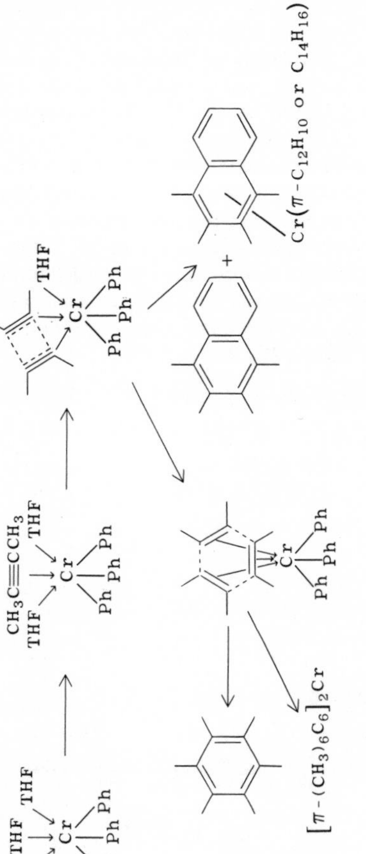
THF
THF
THF
Cr
Ph
Ph
Ph
$CH_3C{\equiv}CCH_3$
THF
THF
Cr
Ph
Ph
Ph
THF
Cr
Ph
Ph
Ph
+
$Cr(\pi\text{-}C_{12}H_{10}$ or $C_{14}H_{16})$
Cr
Ph
Ph
Ph
$[\pi\text{-}(CH_3)_6C_6]_2Cr$

FIGURE 1.3

but-2-ene[67]. The same acetylene is also cyclotrimerised by diphenyl-manganese[109] and by dimesityl derivatives of chromium[97], iron[97], cobalt[98, 110] and nickel[97].

Diphenylacetylene is converted by triphenylchromium into a mixture of hexaphenylbenzene, 1, 2, 3, 4-tetraphenylnaphthalene, 9, 10-diphenyl-phenanthrene, tetraphenylethylene and p-terphenyl[41, 46]. Trivinyl-chromium produces hexaphenylbenzene, 1, 2, 3, 4-tetraphenylbenzene and stilbene[94]. However, tri-o-tolylchromium yields only 1, 2-diphenyl-1, 2-di-o-tolylethylene[46]. Trimethylchromium and diphenyl-acetylene yield hexaphenylbenzene and tetraphenylcyclopentadiene[96]. The latter product is replaced by 1, 2, 3, 4-tetraphenylbenzene when triethylchromium is employed. The powerful hydrogen accepting nature of these organochromium compounds is emphasised by comparison with the organonickel compounds. Thus diphenylacetylene is converted with by diphenylnickel into hexaphenylbenzene only and additionally by diethylnickel into 1, 2, 3, 4-tetraphenylcyclohexa-1, 3-diene[96]. The same acetylene is also cyclotrimerised by dimesityl-chromium[97] and nickel [97, 108].

The only other acetylene examined, dimethyl acetylenedicarboxy-late was converted by triphenylchromium into dimethyl diphenyl-maleate[67].

REFERENCES

1.   American Cyanamid Co., B.P. 901, 326; Chem. Abs., 1963, 59 9932.

2.   Armitage, J. B., U.S.P. 3, 082, 269; Chem. Abs., 1963, 59, 9879.

3.   Arnett, E. M. and Bollinger, J. M., J. Amer. Chem. Soc., 1964, 86, 4729.

4.   Arnett, E. M., Bollinger, J. M. and Sanda, J. C., J. Amer. Chem. Soc., 1965, 87, 2050.

5.   Arnett, E. M. and Strem, M. E., Chem. and Ind., 1961, 2008.

6.   Arnett, E. M., Strem, M. E. and Friedel, R. A., Tetrahedron Letters, 1961, 658.

7.   Azatyan, V. D., Doklady Akad. Nauk. S.S.S.R., 1958, 98, 403.

8.   Azatyan, V. D. and Gyuli-Kevkhyan, R. S., Izvest. Akad. Nauk Armyan S.S.R. khim. Nauki, 1961, 14, 451.

9.   Badger, G. M., Lewis, G. E. and Napier, I. M., J. Chem. Soc., 1960, 2825.

10.   Barborak, J. C., Watts, L. and Pettit, R., J. Amer. Chem. Soc., 1966, 88, 1328.

11. Barnes, C. E., U.S.P. 2, 579, 106; Chem. Abs., 1952, **46**, 6671.

12. Berthelot, M., Compt. rend., 1866, **62**, 905; Ann. Chim. (France), 1866, **9**, 445.

13. Blomquist, A. T. and Maitlis, P. M., J. Amer. Chem. Soc., 1962, **84**, 2329.

14. Booth, G. and Rowe, J. M., Chem. and Ind., 1960, 661.

15. Burt, G. D. and Pettit, R., Chem. Comm., 1965, 517.

16. Canale, A. J. and Kincaid, J. F., U.S.P. 2, 613, 231; Chem. Abs., 1953. **47**, 7547.

17. Chatt, J., Hart, F. A. and Rowe, G. A., B.P. 882, 400; Chem. Abs., 1962, **57**, 12540.

18. Chini, P., De Venuto, G., Salvatori, T. and De Malde, M., Chimica e Industria, 1964, **46**, 1049.

19. Colthup, E. S. and Meriwether, L. S., J. Org. Chem., 1961, **26**, 5169.

20. Cope, A. C. and Campbell, H. C., J. Amer. Chem. Soc., 1952, **74**, 179.

21. Cope, A. C. and Estes, L. L., J. Amer. Chem. Soc., 1950, **72**, 1128.

22. Cope, A. C. and Fenton, S. W., J. Amer. Chem. Soc., 1951, **73**, 1195.

23. Cope, A. C. and Handy, C. T., U.S.P. 2, 950, 334; Chem. Abs., 1961, **55**, 1527.

24. Cope, A. C. and Smith, D. S., J. Amer. Chem. Soc., 1952, **74**, 5136.

25. Craig, L. E. and Larrabee, C. E., J. Amer. Chem. Soc., 1951, **73**, 1191.

26. Dale, J., Chem. Ber., 1961, **94**, 2821.

27. Daniels, W. E., J. Org. Chem., 1964, **29**, 2936.

28. Donda, A. F. and Guerrieri, A., Ricerca sci. Rend., 1964, **6A**, 287.

29. Donda, A. F. and Moretti, G., J. Org. Chem., 1966, **31**, 985.

30. Drefahl, G., Hoerhold, H. H. and Bretschneider, H., J. prakt. Chem., 1964, **25**, 113.

31. Emerson, G. F. Watts, L. and Pettit, R., J. Amer. Chem. Soc., 1965, **87**, 131.

32. Franzus, B., Canterino, P. J. and Wickliffe, R. A., J. Amer. Chem. Soc., 1959, **81**, 1514.

33. Fujisaki, T. and Ogawara, T., Jap. P. 4329('52); Chem. Abs., 1954, **48**, 5214.

34. Fujisaki, T. and Ogawara, T., Jap. P. 5077('56); Chem. Abs., 1958, **52**, 11886.

35. Furlani, A. D., Cervone E. and Biancifiori, M. A., Rec. Trav. chim., 1962, **81**, 585.

36. Green, M. L. H., Nehmé, M. and Wilkinson, G., Chem. and Ind., 1960, 1136.

37. Hagihara, N., J. Chem. Soc. Japan, 1952, **73**. 237.

38. Hagihara, N., J. Chem. Soc. Japan, 1952, **73**, 323.

39. Hagihara, N., J. Chem. Soc. Japan, 1952, **73**, 373.

40. Hardy, D. V. N., B.P. 706, 629; Chem. Abs., 1955, **49**, 9032.

41. Herwig, W., Metlesics, W. and Zeiss, H., J. Amer. Chem. Soc., 1959, **81**, 6203.

42. Higashiura, K., Yokomichi, S. and Oiwa, M., J. Chem. Soc. Japan, Ind. Chem. Sect., 1963, **66**, 379.

43. Hoogzand, C. and Hübel, W., Tetrahedron Letters, 1961, 637.

44. Hoogzand, C. and Hübel, W., Angew. Chem., 1961, 73, 680.

45. Hoover, F. W., Webster, O. W. and Handy, C. T., J. Org. Chem., 1961, **26**, 2234.

46. Huang, Y., Tai, H., Ch'en, S., Hou, K. and Ni, T., Acta Chim. Sinica, 1965, **31**, 149; Chem. Abs., 1965, **63**, 5543.

47. Hübel, W. and Hoogzand, C., Chem. Ber., 1960, **93**, 103.

48. Hübel, W. and Merényi, R., Chem. Ber., 1963, **96**, 930.

49. Hubert, A. J. and Dale, J., J. Chem. Soc., 1965, 3160.

50. Kleinschmidt, R. F. U.S.P. 2, 542, 417; Chem. Abs., 1951, **45**, 7594.

51. Krüerke, U., Hoogzand, C. and Hübel, W., Chem. Ber., 1961, **94**, 2817.

52. Krüerke, U. and Hübel, W., Chem. Ber., 1961, **94**, 2829.

53. Kutepow, N., Berding, C. and Pfab, W., G.P. 1, 102, 141; Chem. Abs., 1962, **56**, 10, 000.

54. Kutepow, N. and Reis, H., G.P. 1, 138, 762; Chem. Abs., 1963, **58**, 10104.

55. Leto, J. R. and Leto, M. F., J. Amer. Chem. Soc., 1961, **83**, 2944.

56. Leto, J. R. and Fiene, M. L., U.S.P. 3, 076, 016; Chem. Abs.,
     1963, **59**, 6276.

57. Longuett-Higgins, H. C. and Orgel, L. E., J. Chem. Soc., 1956,
     1969.

58. Luttinger, L. B., J. Org. Chem., 1962, **27**, 1591.

59. Luttinger, L. B. and Colthup, E. C., J. Org. Chem., 1962, **27**,
     3752.

60. Lutz, E. F., J. Amer. Chem. Soc., 1961, **83**, 2551.

61. McKeever, C. H. and Van Hook, J. O., U.S.P. 2, 542, 551; Chem.
     Abs., 1951, **45**, 7591.

62. Malatesta, L., Santarella, G., Vallarino, L. and Zingales, F.,
     Angew. Chem., 1960, **72**, 34.

63. Meriwether, L. S., Colthup, E. C. and Kennerly, G. W., J. Org.
     Chem., 1961, **26**, 5163.

64. Meriwether, L. S., Colthup, E. C., Kennerly, G. W. and Reusch,
     R. N., J. Org. Chem., 1961, **26**, 5155.

65. Meriwether, L. S. and Fiene, M. L., J. Amer. Chem. Soc., 1959,
     81, 4200.

66. Meriwether, L. S., Leto, M. F., Colthup, E. C. and Kennerly,
     G. W., J. Org. Chem., 1962, **27**, 3930.

67. Metlesics, W. and Zeiss, H., J. Amer. Chem. Soc., 1959, **81**,
     4117.

68. Mills, O. S. and Robinson, G., Proc. Chem. Soc., 1964, 187.

69. Müller, H. and Friederich, H., B.P. 890, 542; Chem. Abs., 1962,
     57, 3357.

70. Murahashi, S., Hagihara, N. and Yamazaki, H., Jap. P. 4814
     ('59); Chem. Abs., 1960, **54**, 14155.

71. Nagasawa, F., Matsusawa, K., Hashizume, G. and Yoshida, K.,
     Jap. P. 3931('55); Chem. Abs., 1957, **51**, 15561.

72. Nagasawa, F. and Yoshida, K., Jap. P. 2428('54); Chem. Abs.,
     1955, **49**, 14803.

73. Overberger, C. G. and Whelan, J. M., J. Org. Chem., 1959, **24**,
     1155.

74. Pirzer, H., and Beck, F., G. P. 1, 092, 908; Chem. Abs., 1962,
     56, 3375.

75. Reikhsfel'd, V. O. and Makovetskii, K. L., Doklady Akad. Nauk S.S.S.R., 1964, **155**, 414.

76. Reikhsfel'd, V. O., Markovetskii, K. L. and Erokhina, L. L., Zhur. obshchei Khim., 1962, **32**, 653.

77. Reppe, W., Pfab, W., Kutepow, N. and Büche, W., G. P. 1, 029, 369; Chem. Abs., 1960, **54**, 22538.

78. Reppe, W., Reicheneder, F., Dury, K. and Suter, H., G. P. 1, 019, 297; Chem. Abs., 1960, **54**, 1365.

79. Reppe, W., Schlichting, O., Klager, K. and Toepel, T., Annalen, 1948, **560**, 1.

80. Reppe, W., Schlichting, O. and Mesiter, H., Annalen, 1948, **560**, 93.

81. Reppe, W., Schlichting, O. and Schweter, W., G. P. 1, 039, 059; Chem. Abs., 1960, **54**, 22538

82. Reppe, W. and Schweckendiek, W. J., Annalen, 1948, **560**, 104.

83. Reppe, W. and Toepel, T., G.P. 859, 464; Chem. Abs., 1956, **50**, 7852.

84. Rose, J. D. and Statham, F. S., J. Chem. Soc., 1950, 69.

85. Sauer, J. C. and Cairns, T. L., J. Amer. Chem. Soc., 1957, **79**, 2659.

86. Schrauzer, G. N., J. Amer. Chem. Soc., 1959, **81**, 5310.

87. Schrauzer, G. N., Chem. Ber., 1961, **94**, 1403.

88. Schrauzer, G. N. and Eichler, S., Chem. Ber., 1962, **95**, 550.

89. Schweckendiek, W., G.P. 1, 072, 244; Chem. Abs., 1961, **55**, 12355

90. Smith, W. R., B.P. 802, 510; Chem. Abs., 1959, **53**, 8070.

91. S.N.A.M. S.p.A., Fr. P. 1, 397, 654; Chem. Abs., 1965, **63**, 8255.

92. Strohmeier, W. and Barbeau, C., Z. Naturforsch., 1964, **19b**, 262.

93. Tanaka, M. and Yamamoto, K., Jap.P. 3819('53); Chem. Abs., 1954, **48**, 8821.

94. Throndsen, H. P., Metlesics, W. and Zeiss, H., J. Organometallic Chem., 1966, **5**, 176.

95. Tsumura, R. and Hagihara, N., Bull. Chem. Soc. Japan, 1964, **37**, 1889.

96. Tsutsui, M. and Zeiss, H., J. Amer. Chem. Soc., 1959, **81**, 6090.

97. Tsutsui, M. and Zeiss, H., J. Amer. Chem. Soc., 1960, **82**, 6255

98. Tsutsui, M. and Zeiss, H., J. Amer. Chem. Soc., 1961, **83**, 825.

99. Ushakov, S. N. and Solomon, O. F., Bull. Acad. Sci. U.S.S.R., 1954, 593.

100. Van Hook, J. O. and Croxall, W. J. U.S.P. 2, 613, 208; Chem. Abs., 1954, **48**, 2772.

101. Watts, L., Fitzpatrick, J. D. and Pettit, R., J. Amer. Chem. Soc., 1965, **87**, 3253.

102. Watts, L., Fitzpatrick, J. D. and Pettit, R., J. Amer. Chem. Soc., 1966, **88**, 623.

103. Wilke, G. and Herrmann, G., Angew. Chem. Internat. Edn., 1962, 1, 549.

104. Wintersberger, K. and Zirger, G., G.P. 1, 025, 870; Chem. Abs., 1960, **54**, 18393.

105. Yamamoto, K. and Oku, M., Bull. Chem. Soc. Japan, 1954, **27**, 382.

106. Yamazaki, H. and Hagihara, N., J. Chem. Soc. Japan, Ind. Chem. Sect., 1958, **61**, 21.

107. Zeiss, H. and Herwig, W., J. Amer. Chem. Soc., 1958, **80**, 2913.

108. Zeiss, H. and Tsutsui, M., U.S.P. 2, 980, 741; Chem. Abs., 1961, **55**, 17583.

109. Zeiss, H. and Tsutsui, M., U.S.P., 3, 122, 567; Chem. Abs., 1964, **60**, 13270.

110. Zeiss, H. and Tsutsui, M., U.S.P. 3, 187, 013; Chem. Abs., 1965, **63**, 9992.

CHAPTER 2

OLIGOMERISATION OF OLEFINS

In view of the relative ease with which acetylenes can be cyclically oligomerised it is surprising that similar reactions have been only cursorily examined for simple olefins. In the course of his early work on the cyclotrimerisation of acetylenes Reppe reported that butyl vinyl ether was cyclotrimerised to isomeric tributoxycyclohexanes in the presence of bis(triphenylphosphine)nickel (II) bromide[73]. Cocyclisation of ethyl acrylate and acetylene using the same catalyst gave ethyl 1, 2-dihydrobenzoate. While this reaction can be envisaged as a direct co-cyclisation of two molecules of acetylene with one of ethyl acrylate an alternative route has come to light. If the same reaction is carried out at lower temperatures using as catalyst bis(triphenylphosphine)nickel cuprocyanide[35], bis(triphenylphosphine)nickel dicarbonyl[17] or bis(acrylonitrile)nickel (0)[78] the product is the ester of hepta-2, 4, 6-trienoic acid. At higher temperatures e.g. 125°C, this is cyclised to a mixture of benzoic and 1, 2-dihydrobenzoic acid esters. Attempts to detect a cyclobutadiene intermediate in this reaction were unsuccessful[10, 76].

Allene shows a close parallelism to acetylene in being converted by a variety of catalysts (Table 2. 1) into a mixture of 1, 2, 4- and 1, 3, 5-trimethylenecyclohexanes (**1** and **2**) and 1, 3, 5, 7-tetramethylene-cyclooctane (**3**)[8, 29, 46, 47]. The amounts of these products formed varies with the catalyst, but the composition of the trimer fraction is practically constant, with 1, 2, 4-trimethylenecyclohexane accounting for about 80% and its 1, 3, 5-isomer for the remainder. It is interesting that, in contrast to the case of acetylene, use of phosphine containing catalysts does not prevent the formation of tetrameric products. As expected 1, 2, 4-trimethylenecyclohexane readily forms Diels-Alder adducts with dienophiles such as methyl acrylate. These adducts are formed _in situ_ if the cyclisation is carried out in the presence of the dienophile[8, 28].

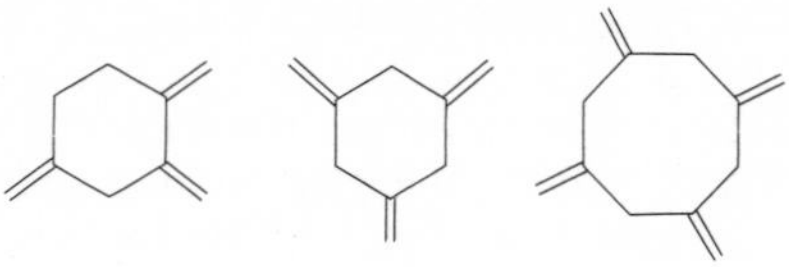

(1)   (2)   (3)

TABLE 2.1

Catalyst	% (1 and 2)	% (3)
$Ni(CO)_2(Ph_3PO)_2$	35	6
$Ni(CO)(Ph_3PO)_3$ ⎱ $Ni(CO)_3(Ph_3PO)$ ⎰	20-25	-
$Ni(CO)_2(Ph_3P)_2$	7	18
$Ni(PCl_3)_4$	24% of 1, 2 and 3	

The discovery that allene has a reactivity comparable to that of acetylene has led to their co-cyclisation[7, 9]. Depending on the catalytic system employed the products are 3,5-dimethylenecyclohexene (4), and either 3,6-dimethylenecyclohexene (5) or 3,5,7-trimethylenecyclooctene (6), cf. Table 2.2. Allene is obviously far

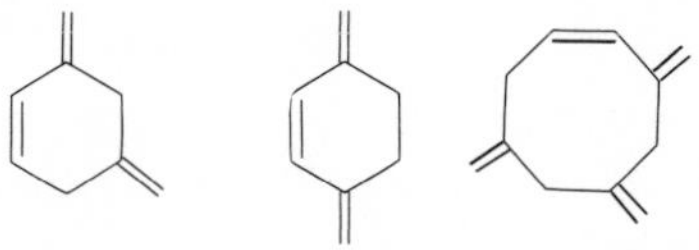

(4)   (5)   (6)

TABLE 2.2

Catalyst	(4)	(5)	(6)
Nickel acetylacetonate	45%	-	5%
Nickel ethylacetoacetate	37%	-	-
Nickel cyanide	45%	-	-
$Ni(CO)_2(Ph_3PO)_2$	17%	26%	-

more reactive than acetylene since not only do the products contain predominantly allene residues but cyclooctatetraene formation does not occur until all of the allene has been consumed. Substituted acetylenes such as propyne, phenylacetylene and vinylacetylene give the appropriate derivatives of (4). In the case of propyne the 1-methyl derivative of (4) predominates.

The very reactive olefin bicyclo[2, 2, 1]hepta-2, 5-diene is converted catalytically into a variety of dimers, the products depending on the catalyst used[3, 11, 12, 39, 45, 79, 85, 88]. The dimers so far identified have the structures (8) to (10). The provisional stereochemistries originally assigned to the stereoisomers of (8) have been corrected subsequently[3], and the structure (7) originally assigned revised to (8c)[17a].

(7)

(8) **a** endo-trans-endo

b exo-trans-exo

c exo-trans-endo

(9)   (10)

As demonstrated by the results in Table 2.3 there is considerable variation in the products formed even with catalysts of the bis(acrylonitrile)nickel (0) type [79]. Bis(acrylonitril)nickel (0) is clearly the most efficient catalyst, functioning at 60°C, where as 1, 1-biphenylene-2, 2-dicyanoethylenenickel (0) and 1, 1, 2-tricyano-1-phenylethylene nickel (0) need temperatures of 80-120°C and give very small yields. The least active of this group of catalysts is bis(fumaronitrile) nickel (0) which gives only traces of dimers at 120°C. It has been pointed out that the catalytic activity thus falls with decreasing electron density at the central nickel atom. The addition of triphenylphosphine markedly increases the overall yield while preventing the formation of (8a or c), 9 and its stereoisomer, and lowering the amount of (8b). Complexing of the catalysts with triphenylphosphine also lowers the electron density on the nickel atom. Comparison of the products obtained using these catalysts with those produced with metal carbonyls (Table 2.4), shows interesting divergencies.

TABLE 2.3

Catalyst*	Overall Yield	Components of dimer mixture in %				
		(8c)	(8b)	(9)	(8a)	stereoisomer of (9)
a	58	60	40	—	—	—
b	2	60	—	—	40	—
c	36	48	52	—	trace	trace
d	29	40	46	9	—	5
e	20	20	10	60	20	—
f	13	32	26	29	—	13
a + Ph_3P	80	66	33	—	—	—
b + Ph_3P	78	98	—	—	trace	—
c + Ph_3P	49	68	30	—	trace	—
d + Ph_3P	45	66	33	—	—	—
e + Ph_3P	45	95	6	—	—	—

*(a) bis(acrylonitrile)nickel (0)

  (b) bis(fumaronitrile)nickel (0)

  (c) 1,1-dicyano-2-phenylethylene nickel(0)

  (d) 1,1-dicyano-2,2-diphenylethylene nickel (0)

  (e) 1,1-biphenylene-2,2-dicyanoethylene nickel (0)

  (f) 1,1,2-tricyano-2-phenylethylene nickel (0)

TABLE 2.4

Catalyst	Products	References
Ni(unsatd. nitrile)$_2$	(8c), (8b), (9), (8a) and stereo-isomer of (9)	79
Ni(CO)$_4$	(8c), (8b)	3, 12
Fe(NO)$_2$(CO)$_2$	(8c), (8b)	39
Co(NO)(CO)$_3$	(8c) trace, (8b), (9)	39
(Ph$_3$P)$_2$Co$_2$(CO)$_6$	(8b), (8c)	3
Fe$_2$(CO)$_9$	(8b), (9), stereoisomer of (9), (10)	11, 45

Transition Metal Intermediates

A different type of dimer, named "Binor-S" is obtained using a binuclear catalyst such as zinc bis-tetracarbonylcobaltate[78a]. This
dimer is obtained exclusively using a high catalyst to monomer ratio.

Addition of Lewis acids, such as boron trifluoride etherate, increases
the yield and rate of formation of "Binor-S". Similar results are obtained with tetracarbonylcobaltates of cadmium (II) and indium (III).
Addition of Lewis bases such as tributylphosphine and 3, 5-lutidine
diverts the reaction to formation of (8b), (8c) and stereoisomers of
(9).

The formation of the hydrocarbon (9) is an example of the more general 2, 6 addition of olefins to bicycloheptadiene catalysed by nickel
catalysts[79, 80, 81]. The preparative value of this reaction is illustrated by the 2, 6-addition of acrylonitrile to bicycloheptadiene.

(11)

Thermal reaction gives only a 12% yield of (11) while bisacrylonitrile nickel (0) catalysis gives a 95% yield. The reaction is also
catalysed by nickel carbonyl, its phosphine derivatives and bis(cy-
cloocta-1, 5-diene)nickel(0) all of which readily undergo ligand exchange with acrylonitrile[81]. In contrast to the 85-95% yields obtained with these catalysts dicobalt octacarbonyl gives only 13% which
is increased to 28% by addition of triphenylphosphine. Iron, manganese and Group VI metal carbonyls as well as chelates of nickel
(II) and other transition metals are ineffective as catalysts. However, bis(triphenylphosphine)nickel cyanide has proved a very active
catalyst giving 80% yields. The use of solvents such as benzene,
acetonitrile, ethanol or pyridine greatly decreases the yield. Analogous 2, 6-adducts of bicycloheptadiene are formed with ethyl acrylate (43%), methacrylonitrile (80%), crotonitrile (85%), methyl maleate (80%) and methyl acetylenedicarboxylate (32%). Diphenylacety-

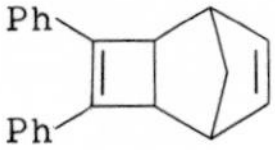

(12)

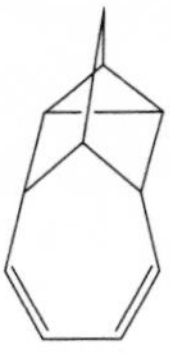

(13)

lene gave the adduct (12) (59%) and acetylene a low yield of (13).
Other acetylenes including phenylacetylene, propargyl alcohol, ethyl
propargylate, hex-1-yne, hept-1-yne, but-2-yne, but-2-yne-1, 4-diol
and its diacetate failed to form adducts. Comparison of the relative
rates of catalytic 2, 6-addition of a series of unsaturated nitriles to
bicyclo[2. 2. 1]heptadiene (B.C.H) with those observed for Diels
Alder addition to 9, 10-dimethylanthracene show marked divergencies, Table 2. 5, which indicate that completely different mechanisms are involved.

TABLE 2. 5

Reaction	Acrylo-nitrile	Croto-nitrile	Methacrylo-nitrile	Me maleate
Cat. addn. to B.C.H.	1·0	0·5	0·3	0·1
D.A. addn. to 9, 10-dimethylanthracene	1·0	0·0075	0·14	0·01

The introduction of organometallic catalysts which mediate the cyclic oligomerisation of buta-1, 3-diene to cycloocta-1, 5-diene and
cyclododeca-1, 5, 9-triene has provided a facile approach to these
alicyclic rings[59, 116]. Although cycloocta-1, 5-diene (14) can be obtained, in addition to 4-vinylcyclohexene (15), by thermal dimerisation the maximum yield obtained is about 15% at 300°C[139, 140].
Cycloocta-1, 5-diene obtained by this method probably arises by a
Cope rearrangement of intermediary cis-1, 2-divinylcyclobutane
(16)[108].

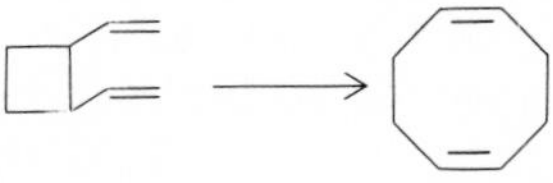

(16)   **(14)**

Transition Metal Intermediates

Three types of catalytic systems can be distinguished on the basis
of the proportions of stereoisomeric cyclododeca-1, 5, 9-trienes produced[15, 116]. The conventional Ziegler system comprising a titanium
halide and an aluminium alkyl converts butadiene into *cis*, *trans*,
trans-cyclododeca-1, 5, 9-triene (**17**) accompanied by small amounts

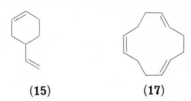

(15)   **(17)**

(~3%) of the all *trans* isomer (**18**). Cycloocta-1, 5-diene and 4-
vinylcyclohexene are also formed in trace amounts. Yields are typically of the order of 90% at 40°C and atmospheric pressure. Although such systems normally promote polymer formation this can
be almost eliminated by using appropriate proportions of titanium
halide and aluminium alkyl. For example, a titanium tetrachloride
and diethylaluminium chloride catalyst with a titanium to aluminium
ratio of 1 to 5 gives exclusively cyclotrimerisation, but lowering the
proportion of diethylaluminium chloride leads to increasing amounts
of polymer formation. It appears that the function of the cocatalyst
is to effect an overall reduction of titanium (IV) to a lower oxidation
state. The necessary reduction can be effected by a variety of alkyl,
aryl, hydride and halogen derivatives of aluminium as well as other
reducing agents such as zinc, aluminium and metal hydrides in conjunction with aluminium chloride, Table 2. 6.

(18)   **(19)**

While titanium (III) is generally implicated as the catalytic species
a titanium (II) complex is also effective. This complex is obtained
from the reaction of titanium tetrachloride with aluminium and aluminium chloride in benzene and has the composition C_6H_6. $TiCl_2$.
$(AlCl_3)_2$[50]. A kinetic study of the cyclisation of butadiene to cyclododecatriene, by titanium tetrachloride and ethylaluminium sesqui-
chloride, shows that the rate determining step is probably the dis-
placement of cyclododecatriene from a titanium π-complex by butadiene[112]. The activity and selectivity of these catalyst systems are

TABLE 2.6 Cyclic Oligomerisation of 1,3-Dienes

1,3-Diene	Catalyst	Products*	References
Buta-1,3-diene	$TiCl_4$ + Et_3Al	Cyclododeca-1,5,9-triene	30, 40
	$TiCl_4$ + Et_2AlCl	Cyclododeca-1,5,9-triene, 85% and cyclohexadeca-1,5,9,13-tetraene	15, 77, 97, 102, 115, 133, 136
	$TiCl_4$ + Et_2AlCl + Me_2SO	C.D.D.T., 90%	63, 77, 87
	$TiCl_4$ + Et_2AlBr	C.D.D.T., 83%	105
	$TiCl_4$ + $Et_3Al_2Cl_6$	C.D.D.T.,	111, 112
	$TiCl_4$ + Et_3Al + Et_2AlCl	C.D.D.T., 90%	52
	$TiCl_4$ + Et_3Al + $EtAlCl_2$	C.D.D.T., 86%	52
	$TiCl_4$ + Et_3Al + $AlCl_3$	C.D.D.T., 87%	4
	$TiCl_4$ + Et_2AlH + $EtAlCl_2$	C.D.D.T., c,t,t 89%; t,t,t - 2.5%	15, 91
	$TiCl_4$ + Et_2AlH + $EtAlCl_2$ + Ph_3P	C.C.C.T., c,l,t 14%; l,t,t - 46%	15
	$TiCl_4$ + Et_2AlH + $AlCl_3$	C.D.D.T.	91
	$TiCl_4$ + i-Bu_2AlH	C.L.D.T., 70%	90
	$TiCl_4$ + i-Bu_2AlH + Et_2AlCl	C.D.D.T., 84%	52
	$TiCl_4$ + Ph_3Al	C.D.D.T.	127
	$TiCl_4$ + $Ph_3Al_2Cl_3$	C.D.D.T.	127
	$TiCl_4$ + Al + $Et AlCl_2$	C.D.D.T.	130
	$TiCl_4$ + Al + $AlCl_3$	C.D.D.T., 70%	55, 131
	$TiCl_4$ + Al + $AlCl_3$ + Ph_2O	C.D.D.T., 89%	54
	$TiCl_4$ + Al + $AlCl_3$ + Ph_2S	C.D.D.T., 96%	131

(*Continued overleaf*)

37

Table 2.6 (continued)

1,3-Diene	Catalyst	Products*	References
Buta-1,3-diene	$TiCl_4 + Al + AlCl_3 + NaCl$	C.D.D.T., 82%	131
	$TiCl_4 + Al + AlCl_3 + Et_2AlCl$	C.D.D.T., 87%	131
	$TiCl_4 + Al + AlCl_3 + Et_3Al$	C.D.D.T., 90%	131
	$TiCl_4 + Al + AlCl_3 + Et_2AlH$	C.D.D.T., 94%	131
	$TiCl_4 + Zn + AlCl_3$	C.D.D.T.	56
	$TiCl_4 + AlH_3$	C.D.D.T.	91
	$TiCl_4 + LiAlH_4$	C.D.D.T.	91
	$TiCl_4 + LiAlH_4 + AlCl_3$	C.D.D.T.	91
	$TiCl_4 + NaH + AlCl_3$	C.D.D.T.	91
	$TiCl_4 + CaH_2 + AlCl_3$	C.D.D.T., 82%	15, 91
	$TiCl_4 + MgBr + AlCl_3 + dithian$	C.D.D.T.	62
	$Ti(OEt)_4 + Et_2AlCl$	C.D.D.T., 81%	97
	$Ti(OEt)_4 + Ph_2Al_2Cl_4$	C.D.D.T.	127
	$Ti(OEt)_4 + Al + AlCl_3$	C.D.D.T.	55
	$Ti(OBu)_4 + Et_2AlCl$	C.D.D.T., 87%	18
	$Ti(OBz)_4 + Et_2AlCl$	C.D.D.T., 71%	134
	$TiCl_3 + Et_2AlCl$	C.D.D.T.	30
	$TiCl_3 + Ph_3Al_2Cl_3$	C.D.D.T.	127
	$TiCl_3NEt_2 + Et_2AlCl$	C.D.D.T.	103

38

Catalyst	Products	Ref.
$TiCl_3(\nu\text{-}C_8H_{17}NH_2) + Et_2AlCl$	C.D.D.T., 87%	49
$PhTiAlCl_8 + Ph_3Al_2Cl_3$	C.D.D.T., 83%	127
C.D.D.T. Ni(0)	C.D.D.T.	122
$(\pi\text{-allyl})_2Ni$	C.D.D.T.	117
$(C.O.D.)Ni(P\,Ph_3)_2$	C.O.D.	125
$(C.O.D.)_2Ni + (PhO)_3P$	V.C.H., 8%; C.O.D, 77.5%; C.D.D.T., 2%	93
$(C.O.D.)Ni(duroquinone) + (PhO)_3P$	V.C.H. and C.O.D.	137
$(acrylonitrile)_2Ni + (RO)_3P$		
R = phenyl	V.C.H., 9%; C.O.D, 26.5%	137
R = p-cresyl	C.O.D, 67%	5
R = guiacyl	C.O.D, 77%	5
R = diphenyl	C.O.D, 84%	5
R = thymyl	C.O.D, 86%	5
$Ni(PPh_3)_4$	C.O.D.	125
$Ni(PPh_3)_2$	C.O.D.	125
$Ni[(PhO)_3P]_3$	V.C.H., 10%; C.O.D, 51%	42
$Ni[(o\text{-tolyl-}O)_3P]_3$	V.C.H., 7%; C.O.D, 83%	42, 60
$Ni[(3,5\text{-dimethylphenyl-}O)_3P]_3$	V.C.H., 11%; C.O.D, 62%	42
$Ni(CO)_4$	C.O.D, 23%; C.D.D.T., 69%	58
$Ni(CO)_2(Ph_3P)_2$	V.C.H., 9%; C.O.D, 37.5%	70, 71
$Ni(CO)_2(Ph_3Sb)_2$	C.O.D, 34%; V.C.H., C.D.D.T.	84

(Continued overleaf)

Table 2.6 (continued)

1,3-Diene	Catalyst	Products*	References
Buta-1,3-diene	Ni(CO)$_2$[(PhO)$_3$P]$_2$	C.O.D., 78%; C.D.D.T., 14%	13, 16, 21, 48, 70, 72, 83, 105
	Ni(CO)[(PhO)$_3$P]$_3$	V.C.H.; C.O.D.	21, 48
	Ni(CO)$_2$[(MeO)$_3$P]$_2$	V.C.H.; C.O.D.	13, 70
	Ni(CO)$_2$(Ph$_3$PO)$_2$	C.O.D.	14
	Ni(CO)$_4$ + Et$_3$Al	V.C.H.; C.O.D.; C.D.D.T.	22
	(π-allyl) (π-C$_5$H$_5$)Ni + Ph$_3$P + Et$_3$Al	V.C.H., 8%; C.O.D., 47%; C.D.D.T., 3%	100
	(π-C$_5$H$_5$)$_2$Ni picrate + Ph$_3$P + Et$_3$Al	V.C.H., 16%; C.O.D., 77%; C.D.D.T., 5%	101
	NiCl$_2$ + PhMgCl + (o-anisyl-O)$_3$P	C.O.D.	129
	NiCO$_3$ + i-Bu$_2$AlH + (thymyl-O)$_3$P	C.O.D.	129
	Ni acetate + Et$_2$AlCl + (o-PhOC$_6$H$_4$O)$_3$P	C.O.D.	129
	Ni 2-ethylhexanoate + Et$_3$Al + Ph$_3$P	V.C.H., 21%; C.O.D., 74%	43
	Ni(acac)$_2$ + Et$_2$AlOEt	C.D.D.T.; l,l,l, 65%, c,l,l, -7%, c,c,l - 9%	15
	Ni(acac)$_2$ + Al + Ph$_3$P	V.C.H., 9%; C.O.D., 85%; C.D.D.T., 1%	57
	Ni(acac)$_2$ + Et$_2$AlOEt + Ph$_3$P	V.C.H., 21%; C.O.D., 67%; C.D.D.T., 10%	92, 124
	Ni(acac)$_2$ + Et$_3$Al + (o-tolyl-O)$_3$P	C.O.D.	129
	Ni(acac)$_2$ + Et$_2$AlOEt + (o-tolyl-O)$_3$P	C.O.D., 92%	129
	Ni(acac)$_2$ + Et$_3$Al + (2,4-xylyl-O)$_3$P	C.O.D.	129

Catalyst	Products	Ref.
$Ni(acac)_2 + Et_3Al + (o\text{-}PhC_6H_4O)_3P$	C.O.D.	129
$Ni(acac)_2 + Et_2AlOEt + (\alpha\text{-naphthyl-}O)_3P$	V.C.H., 7%; C.O.D., 80%; C.D.D.T., 0.4%	93
$Ni(acac)_2 + LiAlH_4 + (\beta\text{-naphthyl-}O)_3P$	C.O.D.	129
Ni dimethylglyoxime + BuLi + $(o\text{-}ClC_6H_4O)_3P$	C.O.D.	129
$NiEt_2$.bipyridyl	C.D.D.T.	132
$Ni(CN)_2$	C.O.D., 1.5%	34, 70
$FeEt_2(bipyridyl)_2$	V.C.H., 23%; C.O.D., 72%	132
$(\pi\text{-allyl})_2CrI$	C.D.D.T., 40-70%	118
$CrCl_3 + Et_3Al$	C.D.D.T., 75%	102
$CrCl_3 + i\text{-}Bu_3Al$	C.D.D.T., 86%	102, 136
$CrO_3 + AcCl + Et_3Al$	C.D.D.T., 87%; V.C.H. and C.O.D.1%	53
$CrO_2Cl_2 + AlH_3$	C.D.D.T. t,t,t - 35%, c,t,t - 15%	91
$CrO_2Cl_2 + Et_3Al$	C.D.D.T. t,t,t - 54%. c,t,t - 36%	121
$CrO_2Cl_2 + Et_2AlOEt$	C.D.D.T. t,t,t - 47%, $c.t.t$ - 33%	15
$TiCl_4 + Et_3Al$ (Isoprene)	Trimethylcyclododeca-1,5,9-triene	30
$TiCl_4 + i\text{-}Bu_3Al$	2,6-Dimethylocta-1,3,6-triene and other dimers and trimers	135
$TiCl_4 + Et_2AlCl$	Trimethylcyclododeca-1,5,9-triene	115
$TiCl_4 + Et_3Al + Et_2AlCl$	Trimethylcyclododeca-1,5,9-triene, 60%	52
$TiCl_4 + Al + AlCl_3$	Trimethylcyclododeca-1,5,9-triene	131

(Continued overleaf)

Table 2.6 (continued)

1,3-Diene	Catalyst	Products*	References
Isoprene	$TiCl_4 + Zn + AlCl_3$	Trimethylcyclododeca-1,5,9-triene	56
	$Ti(OEt)_4 + Al + AlCl_3$	Trimethylcyclododeca-1,5,9-triene	55
	$Ti(OBu)_4 + Et_2AlCl$	19% 2,3-dimethyl-1,3,6-octatriene + 33% cyclic and linear trimers + 11% cyclic tetramer	18,98
	$Ti(OPh)_4 + Et_2AlCl$	30% 2,3-dimethyl-1,3,6-octatriene + 18% cyclic and linear trimers + 5% cyclic tetramers	98
	$TiCl_3 + Et_2AlCl$	1,5,9-cyclododecatrienes	30
	$Ni(CO)_2[(PhO)_3P]_2$	Dimethylcyclooctadiene	16
	$Ni(acac)_2 + Et_2AlOEt + Ph_3P$	p-Diprene 14%, dipentene 5%, 1,5-dimethylcyclooctadiene, 61%, 1,5,9-trimethylclododecatriene	124
	$Ni(acac)_2 + Al + Ph_3P$	Dimer and Trimer	57
	$NiBr_2(Ph_3P)_2 + LiBH_4$	49% Dimers comprising 57.5% dimethylocta-1,5-diene, 28% 2,4-dimethyl-4-vinylcyclohexene and 14.5% methylisopropenylcyclohexene and 43% trimers	138
	$Ni(acac)_2 + (PhO)_3P + LiBH_4$	Above dimers in 78%, 13% and 9% respectively	138
	$Ni(acac)_2 + (o\text{-tolyl-O})P + LiBH_4$	Above dimers in 63%, 22% and 15% respectively	138

Monomer	Catalyst	Product	Reference
Piperylene (penta-1,3-diene)	$CrO_2Cl_2 + AlH_3$	Trimethylcyclododecatrienes	91, 123
	$TiCl_4 + Et_3Al$	Trimethylcyclododecatriene	30
	$TiCl_4 + Zn + AlCl_3$	Trimethylcyclododecatriene	56
	$Ti(OBu)_4 + Et_2AlCl$	Trimethylcyclododecatriene	18
	$TiCl_3 + Et_2AlCl$	Trimethylcyclododecatriene	30
	$Ni(acac)_2 + Et_2AlOEt + Ph_3P$	Dimethylcyclooctadiene, 92%	124
	$CrO_2Cl_2 + AlH_3$	Trimethylcyclododecatriene	91, 123
	$(\pi\text{-}C_5H_5)(\pi\text{-}PhC_6H_5)Mn$	3,7 and 3,4-Dimethylcyclooctadienes	69
2,3-Dimethylbuta-1,3-diene	$TiCl_4 + Zn + AlCl_3$	Trimer	56
	$Ti(OEt)_4 + Al + AlCl_3$	Trimer	55
	$PhTiAlCl_8 + Ph_3Al_2Cl_3$	Trimer	127
	$NiCl_2 + Mg + Ph_3P$	Tetramethylcyclooctadiene	57
	$CrO_2Cl_2 + Et_3Al$	Linear trimer 68% and a tetramer	114
Cyclohexa-1,3-diene	$TiCl_4 + Zn + AlCl_3$	Trimer	56
	$Ti(OEt)_4 + Al + AlCl_3$	Trimer	55
	$PhTiAlCl_8 + Ph_3Al_2Cl_3$	Cyclic dimer	127
Chloroprene	$Ni(CO)_2[PhO)_3P]_2$	Dichlorocyclooctadiene	16
	$Ni(CN)_2$	Dichlorocyclooctadiene	34
2,3-dichlorobuta-1,3-diene	$Ni(CN)_2$	Tetrachlorocyclooctadiene 12%	34

* C.D.D.T. = cyclododeca-1,5,9-triene; C.O.D. = cycloocta-1,5-diene; V.C.H. = 4-vinylcyclohexene.

improved by the addition of substances such as sodium chloride or diphenylsulphide which complex with any aluminium chloride present[131]. It is noteworthy that addition of triphenylphosphine causes the all-*trans*-cyclododecatriene (**18**) to preponderate (~46%) over the *cis*, *trans*, *trans*-isomer (17) (~14%)[15].

Catalysts based on nickel(0) in the form of bisphosphine and bisphosphite nickel dicarbonyl complexes were the first to be found to catalyse the cyclisation of butadiene to cycloocta-1, 5-diene and 4-vinylcyclohexene with only small amounts of cyclododeca-1, 5, 9-triene being detectable [70]. These catalysts generally need activating by pretreatment with acetylene[70] or olefins[16, 83]. The use of catalysts derived by reaction of nickel acetylacetonate, formate or dimethylglyoximate with diethylaluminium ethoxide produces all-*trans* (~65%), *cis*, *trans*, *trans* (~7%) and *cis*, *cis*, *trans* (~9%) isomers of cyclododeca-1, 5, 9-triene with minor quantities of cyclo-octa-1, 5-diene and 4-vinylcyclohexene[15]. Reaction of nickel acetylacetonate with triethylaluminium in the presence of α, α'-dipyridyl gives the crystalline α, α'-dipyridylnickeldiethyl which also catalyses the cyclotrimerisation[132]. Other complexes of nickel(0) are equally effective, (Table 2. 6). The presence of a phosphine or a phosphite reduces the yield of cyclododecatrienes to minor amounts (~1%); cyclocta-1, 5-diene (80-90%) becomes the major product with lesser amounts of 4-vinylcyclohexene. In the case of phosphites the proportionate yield of cyclooctadiene increases markedly with the size of the aryl group.

The key to elucidating the mechanism of cyclisation reactions with nickel(0) was provided by the discovery that olefin nickel(0) complexes can be obtained by reacting nickel acetylacetonate with diethylaluminium ethoxide in the presence of an olefin[120]. In this way complexes such as bis(cycloocta-1, 5-diene)nickel(0) (**20**) and the coordinatively unsaturated cyclododeca-1, 5, 9-triene nickel(0) (**21**) have been prepared. Some of the more relevant reactions of the latter complex are depicted in Figure 2. 1[116, 122]. This deep red crystalline complex is monomeric in the vapour phase and in solution. It is air sensitive but stable to water. Solutions of the complex react rapidly with hydrogen forming cyclododecane and nickel. Being coordinatively unsaturated complex (**21**) forms stable 1:1 complexes (**22**) with phosphines. Carbon monoxide also forms a 1:1 complex (**23**) at −80°C, which decomposes above 0°C. The cyclododecatriene nickel (0) complex undergoes facile ligand replacement reactions. For example cycloocta-1, 5-diene replaces the cyclododecatriene forming the complex (**20**). At 20°C cyclododecatriene nickel(0) reacts with butadiene in a smooth homogeneous reaction producing cyclododeca-1, 5, 9-triene. The same reaction is observed with bis(cycloocta-1, 5-diene) nickel (0). At −40°C complex (**21**) reacts with 3 moles of butadiene liberating cyclododecatriene. A new crystalline complex

(24) is produced, which is also capable of catalysing the formation of cyclododecatriene.

FIGURE 2.1

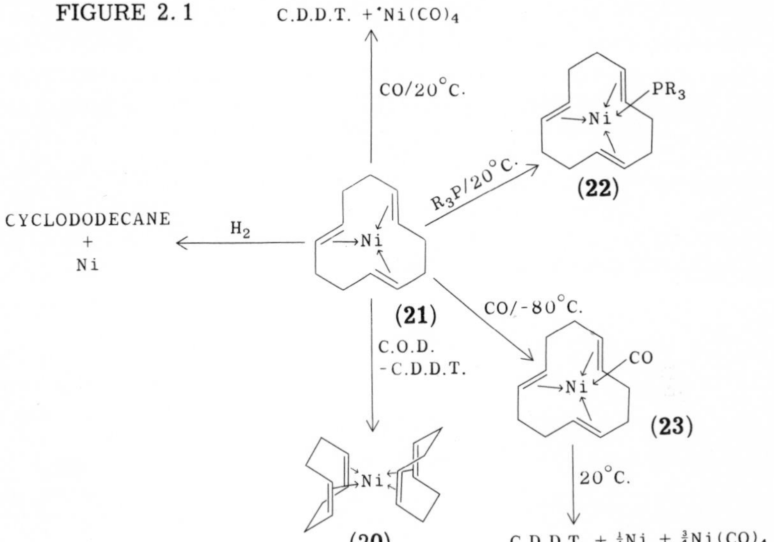

The principal reactions of complex (24) are shown in Figure 2.2. Hydrogenation gives n-dodecane in quantitative yield. Heating produces cyclododecatriene and so does reaction with phosphines, carbon monoxide and butadiene. Under mild conditions reaction with phosphines produces the same complex (22) as is formed by cyclododecatriene nickel(0) (21). Surprisingly the reaction with carbon

FIGURE 2.2

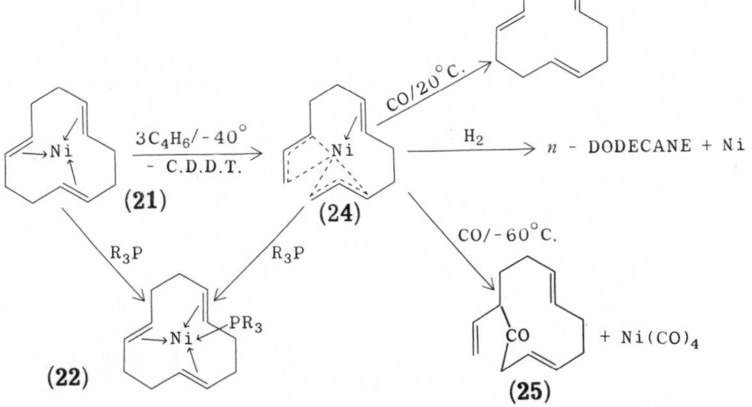

monoxide at low temperatures results in the formation of 2-vinyl-cycloundecanone (25). Completely analogous reactions are observed with bis(π-allyl)nickel, which also catalyses the cyclotrimerisation of butadiene[117].

The foregoing observations indicate that cyclododecatriene formation involves firstly the replacement of weak ligands coordinated to nickel(0) by butadiene. This is followed by linkage of the three butadiene molecules to give complex (24). Finally (24) is converted at the reaction temperature into (21). The catalytic cycle is completed by release of cyclododecatriene from (21) by reaction with butadiene. This reaction sequence suggests that for formation of cyclooctadiene it is necessary to block one of the coordination centres with a ligand not readily displaced by butadiene.

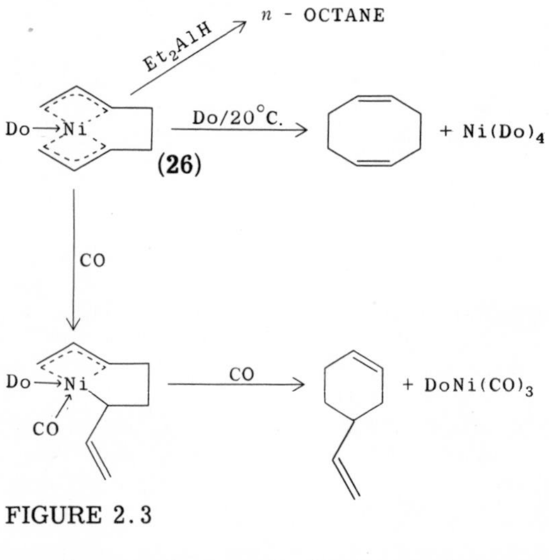

FIGURE 2.3

Weak donors such as arsines are ineffective and success was first obtained by reduction of the nickel compound in the presence of phenylacetylene giving a red catalyst solution which converted butadiene at 80°C into 4-vinyl-cyclohexene (8%), cyclooctadiene (24%) and cyclododecatriene (63%). Far more cyclooctadiene (70%) and 4-vinylcyclohexene (20%) and less trimer (10%) are obtained when triphenylphosphine replaces the phenylacetylene. The same result is obtained using the well-defined tetrakis(triphenylphosphine)nickel (0)[125]. Using tri(2-diphenyl)phosphite as a donor complex (26) has been isolated at low temperature[116]. Warming to 20°C and adding a donor such as triphenylphosphine liberates cyclooctadiene, whereas low temperature reduction with diethylaluminiumhydride gives n-octane instead. The course of the catalytic formation of 4-vinylcyclohexene has also been clarified. With some catalysts this is the major product. An illustration of this is supplied by carbon monoxide which presumably first disrupts one of the two π-allyl systems, cf. Figure 2.3. Subsequent formation of a carbon to carbon bond and detachment from the nickel produces 4-vinylcyclohexene. This accounts for the adverse effect of carbon monoxide on these cyclisation reactions. As might have been predicted it appears that both cyclooctadiene and 4-vinylcyclohexene have the same precursor, with the electronic and steric factors of the donor determining the relative amounts of the two products.

An extension of these reactions is the co-cyclotrimerisation of butadiene with ethylene. Using a nickel catalyst suitable for cyclooctadiene synthesis both cyclooctadiene and *cis, trans*-cyclodeca-1, 5-diene (**27**) are obtained[60, 116, 119]. The latter compound is formed in 80% yield to the exclusion of cyclododecatriene on an unblocked nickel(0) catalyst. The cyclododecadiene is rapidly transformed above 80°C into 1, 2-divinylcyclohexane (**28**) via a Cope rearrangement.

(**27**)   (**28**)

Analogously, 1, 2-dimethyl-cyclodeca-1, 4, 8-triene (**29**) is obtained from but-2-yne and butadiene[116]. This compound can be thermally rearranged to (**30**).

(**29**)   (**30**)

The formation of *all trans*-cyclododeca-1, 5, 9-triene when triphenylphosphine is added to a Ziegler type catalyst was mentioned on page 44. The *all trans* isomer ($\sim 47\%$) also preponderates over *cis, trans, trans*-cyclododecatriene ($\sim 33\%$) when a catalyst derived from reaction of an alkylaluminium with chromium trioxide[53], chloride[102, 123, 136] or chromyl chloride[15, 91, 121] is used. As this catalyst reacts with but-2-yne forming the bis(hexamethylbenzene)chromium cation it has been suggested that the catalyst contains monovalent chromium[121, 125]. However, in view of the reactions of trialkylchromiums surveyed in Chapter 1, this conclusion is probably incorrect. Triallylchromium polymerises butadiene to 1, 2-polybutadiene but diallylchromium iodide converts butadiene into *all trans*-and *cis,trans,trans*-cyclododecatrienes[118]. The use of polar solvents such as chlorobenzene or dichloromethane which favour dissociation of the dimeric diallylchromium halides considerably increases the yield. The same effect is observed when aluminium halides or alkyl-aluminium halides are added again presumably as a result of promoting dissociation; in this case by formation of the complex $(C_3H_5)_2CrAlX_4$. The same isomers of cyclododecatriene are also produced when a catalyst obtained from reaction of ferric chloride[56] or acetylacetonate[125] with an aluminium alkyl is used. In this case however, appreciable amounts of open chain oligomers are also produced. Reaction of ferric trisacetylacetonate with triethylaluminium in the presence of α,α'-dipyridyl produces the crystalline complex (31) which converts butadiene into cycloocta-1,5-diene (72%) and 4-vinylcyclohexene (23%). Ethane and ethylene are evolved. Dimerisation of 1,1,4,4-tetradeuterobutadiene gave products without any randomnisation of the deuterium[132].

(31)

The variations in stereochemistry of the cyclododecatrienes depending on whether titanium, nickel or chromium catalysts are used must reflect different steric situations on these metals. These variations are of considerable practical utility as interconversion of the stereoisomeric cyclododeca-1,5,9-trienes is not readily effected[116]. Interconversion of the *all trans* and the *cis,trans,trans* isomers has been achieved using chromium trisacetylacetonate and triethyl-aluminium at 200°C but the main products are of unknown structure. The *all trans*-cyclododecatriene appears to be the more thermally stable isomer. At 500°C both stereoisomers are non-catalytically converted into the four isomeric 1,2,4-trivinylcyclohexanes.

The catalytic cyclisation of butadiene has been extended to isoprene, piperylene, 2,3-dimethylbutadiene, chloroprene and cyclohexa-1,3-diene, Table 2.6. Depending on the catalyst system employed stereoisomeric cyclododeca-1,5,9-trienes or cycloocta-1,5-dienes have been isolated but not throughly investigated. In some cases open chain oligomers are formed[113, 114, 135] and this obviously casts doubts on the structures attributed to the products reported.

As mentioned earlier the catalyst derived from iron trisacetyl-acetonate and triethylaluminium converts butadiene into *n*-dodeca-1,

3, 6, 10-tetraene, 3-methylhepta-1, 4, 6-triene, cycloocta-1, 5-diene, cyclododeca-1, 5, 9-trienes and 4-vinylcyclohexene[38, 125]. Optimum conditions are realised when triethylaluminium is added to the iron trisacetylacetonate in the presence of butadiene[38]. The optimum aluminium to iron ratio is 3:1. If the reaction is carried out in the presence of ethylene hexa-1, 4-diene is formed[104]. A range of 1, 4-dienes have been obtained by condensing ethylene with substituted butadienes (Table 2. 7). The following examples indicate that it is the more highly substituted double-bond of the 1, 3-diene which reacts preferentially (Figure 2. 4).

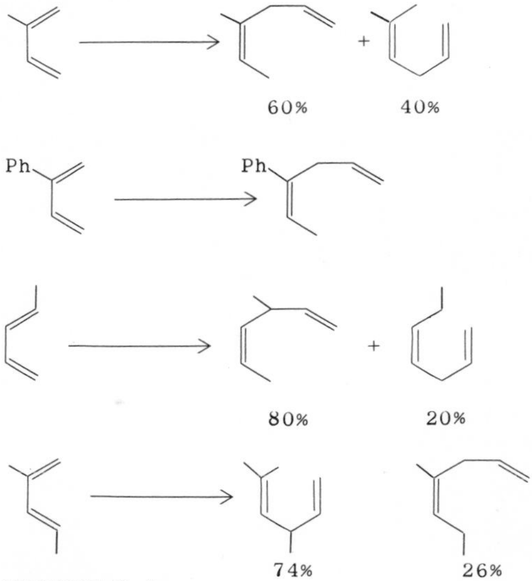

FIGURE 2. 4

Cobalt catalysts obtained from the reaction of triethylaluminium with cobalt chloride[99], cobalt acetylacetonate[74, 125, 126, 128] or dicobalt octacarbonyl[65, 66] convert butadiene into 3-methylhepta-1, *trans* 4, 6-triene and a little octa-1, 3, 6-triene resulting from 1, 2 and 1, 4-addition. Experiments with 1, 1, 4, 4-tetradeuterobutadiene show that only the hydrogen atoms attached to carbon atoms 1 and 4 undergo migration, i.e.

$$D_2C = CH.CH = CD_2 \longrightarrow D_2C = CH - \overset{\overset{\displaystyle CD_3}{\displaystyle |}}{CH} - CD = CH - CH = CD_2$$

In contrast to the iron catalysts just discussed and analogous nickel catalysts (vide infra) two of the double bonds in these products are conjugated, (Table 2. 7). The term dienylation has been applied to this type of reaction[128].

TABLE 2.7 Linear Oligomerisation of Olefins

Olefin(s)	Catalyst	Products	References
Ethylene	$(\pi\text{-}C_5H_5)TiCl_3 + Na/Hg$	But-1-ene and hex-1-ene	89
	Ni palmitate + Et_3Al + PhC_2H	Butenes	67
	$Ni(acac)_2 + Et_3Al$	But-1-ene and some but-2-ene and hex-1 and-2-enes	36
	$Co(acac)_2 + Et_3Al$	Butene	37
	$RhCl_3$	Butenes	2
	$RuCl_3$	Butenes	2
	$PdCl_2$	Butenes	23, 86
	$PdCl_2(C_2H_4)$	Butenes	106
Propylene	$(\pi\text{-allyl})$ NiI + $AlBr_3$ + Ph_3P	n-Hexene, 20% and 2-methylpentene 75%	94
	$(\pi\text{-allyl})$ NiI + $AlBr_3$ + $(C_6H_{11})_3P$	2,3-Dimethylbutene 67% and 2-methyl-pentene 31%	94
But-1-ene	$RhCl_3$	C_6 olefins, mainly 2-methylpent-2-ene	2
	$CH_3TiCl_3 + CH_3AlCl_2$	42% dimer (mainly 2-ethylhex-1-ene) and 19% trimers	9a
Propylene + ethylene	$RhCl_3$	n-Pentene, 2-methylbutene, n-hexene, 2-methylpentene	2
Butadiene	$Ni(acrylonitrile)_2 + (EtO)_3P$	Octa-1,3,6-triene 57%, octa-1,3,7-triene 5%	59, 82
	$Ni(acrylonitrile)_2 + Bu_3P$	Octa-1,3,7-triene	59
	$Ni(acrylonitrile)_2 + (morpholide)_3P$	Octa-2,4,6-triene	59

	Catalyst	Product	References
	$Co(allyl)_3$	*trans*-3-Methylhepta-1,4,6-triene	94
	$CoCl_2 + NaBH_4$	3-Methylhepta-1,4,6-triene	61
	$CoCl_2 + Et_3Al$	Octatriene 80%	99
	$Co(acac)_3 + Et_3Al$	*trans*-3-methylhepta-1,4,6-triene + octa-1,3,6-triene	51,74 126,128
	$Co_2(CO)_8 + Et_3Al$	*trans*-3-Methylhepta-1,4,6-triene	65,66
	$Fe(acac)_3 + Et_3Al$	*n*-Dodeca-1,3,6,10-tetraene, 3-methylhepta-1,4,6-triene	38
	$FeCl_3 + Et_3Al + Ph_3P$	Octa-1,3,*cis* 6-triene, 59%; 3-methyl-hepta-1,4,6-triene 26%	96
	$FeCl_3 + Et_3Al + Ph_3P$	Octa-1,3-*trans* 6-triene 28%; 3-methyl-hepta-1,4,6-triene 31%	96
	$RhCl_3$	Octa-2,4,6-triene (isomers), unconjugated octatriene	1,2
Butadiene + ethylene	$Ni(acac)_2 + Et_3Al + (o\text{-tolyl } O)_3P$	Deca-1,4,9-triene	59
	$NiCl_2(Bu_3P)_2 + i\text{-}Bu_2AlCl$	Hexa-1,4-diene	41
	$Co(acac)_2 + Et_3Al$	Hexa-1,3-diene	75,126,128
	$Fe(acac)_3 + Et_2AlOEt$	Hexa-1,4-diene	104
	$RhCl_3$	1,4 and 2,4-Hexadiene, 3-ethylhexa-1,4-diene, 3-methylhepta-1,4-diene	2,107
Butadiene + propylene	$NiCl_2(Bu_3P)_2 + i\text{-}Bu_2AlCl$	2-Methylhexa-1,4-diene	41
	$RhCl_3$	*trans* + little *cis*-2-Methylhexa-1,4-diene	2
Pent-1-ene	$CH_3TiCl_3 + CH_3AlCl_2$	67% Dimers (mainly 2-*n*-propylhept-1-ene) and 18% trimers	9a

(Continued overleaf)

Olefin(s)	Catalyst	Products	References
Penta-1,3-diene + ethylene	$NiCl_2(Bu_3P)_2 + i\text{-}Bu_2AlCl$	3-Methylhexa-1,4-diene	41
	$Fe(acac)_3 + Et_2AlOEt$	3-Methylhexa-1,4-diene and hepta-1,4-diene	104
	$RhCl_3$	3-Methylhexa-1,4-diene	2
Isoprene + ethylene	$NiCl_2(Bu_3P)_2 + i\text{-}Bu_2AlCl$	4-Methylhexa-1,4-diene	41
	$Fe(acac)_3 + Et_2AlOEt$	4-Methylhexa-1,4-diene 60%, 5-methylhexa-1,4-diene 40%	104
	$RhCl_3$	4-Methylhexa-1,4-diene	2
Isoprene + propylene	$NiCl_2(Bu_3P)_2 + i\text{-}Bu_2AlCl$	2,4-Dimethylhexa-1,4-diene	41
	$RhCl_3$	2,4-Dimethylhexa-1,4-diene	2
Hex-1-ene	$CH_3TiCl_3 \quad CH_3AlCl_2$	65% Dimers (mainly 2-n-butyloct-1-ene) and 19% trimers	9a
Hex-1-ene + butadiene	$NiCl_2(Bu_3P)_2 + i\text{-}Bu_2AlCl$	2-Butylhexa-1,4-diene	41
Hexa-1,3-diene + ethylene	$Fe(acac)_3 + Et_2AlOEt$	3-Ethylhexa-1,4-diene, octa-1,4-diene	104
Hexa-2,4-diene	$NiCl_2(Bu_3P)_2 + i\text{-}Bu_2AlCl$	3-Ethylhexa-1,4-diene, 3-methylhepta-1,4-diene	41
2-Methylpenta-1,3-diene + ethylene	$Fe(acac)_3 + Et_2AlOEt$	3,5-Dimethylhexa-1,4-diene, 4-methylhepta-1,4-diene	104
2-Ethylbuta-1,3-diene + ethylene	$Fe(acac)_3 + Et_2AlOEt$	4- and 5-Ethylhexa-1,4-diene	104

2,3-Dimethylbuta-1,3-diene + ethylene	$Fe(acac)_3 + Et_2AlOEt$	4,5-Dimethylhexa-1,4-diene, 2,3-dimethylhexa-2,4-diene	104
Cyclohexa-1,3-diene + ethylene	$NiCl_2(Bu_3P)_2 + i\text{-}Bu_2AlCl$	3-Vinyl cyclohexene	41
Hept-1-ene	$CH_3TiCl_3 + CH_3AlCl_2$	63% dimers (mainly 2-n-pentylnon-1-ene) and 17% trimers	9a
Oct-1-ene	$CH_3TiCl_3 + CH_3AlCl_2$	62% Dimers (mainly 2-n-hexyldec-1-ene)	9a
Octa-1,3-diene + ethylene	$Fe(acac)_3 + Et_2AlOEt$	3-Butylhexa-1,4-diene, deca-1,4-diene	104
Cycloocta-1,3-diene + ethylene	$NiCl_2(Bu_3P)_2 + i\text{-}Bu_2AlCl$	3-Vinylcyclooctene	41
2-Pentylbuta-1,3-diene + ethylene	$Fe(acac)_3 + Et_2AlOEt$	4 and 5-Pentylhexa-1,4-dienes	104
Styrene + ethylene	$RhCl_3$	2-Phenylbut-2-ene	2
Styrene + butadiene	$Ni(acac)_2 + Et_3Al + (o\text{-}tolyl\text{-}O)_3P$	Phenyldeca-1,4,8-triene	44
Styrene + isoprene	$Ni(acac)_2 + Et_3Al + (o\text{-}tolyl\text{-}O)_3P$	1-Phenyl-4,8-dimethyldeca-1,4,8-triene	59
Phenylacetylene + butadiene	$Co(acac)_2 + Et_3Al$	1,8-Diphenylocta-1,3,5,7-tetraene	126,128
α-Methylstyrene + butadiene	$Ni(acac)_2 + Et_3Al + (o\text{-}tolyl\text{-}O)_3P$	2-Phenylundecatriene	44
β-Methylstyrene + butadiene	$Ni(acac)_2 + Et_3Al + (o\text{-}tolyl\text{-}O)_3P$	1-Phenyl-2-methyldecatriene	44
2-Phenylbutadiene + ethylene	$Fe(acac)_3 + Et_2AlOEt$	4-Phenylhexa-1,4-diene	104
Divinylbenzene + butadiene	$Ni(acac)_2 + Et_3Al + (o\text{-}tolyl\text{-}O)_3P$	Didecatrienyl benzene, decatrienylvinyl-benzene	44,59

(Continued overleaf)

TABLE 2.7 (continued)

Olefin(s)	Catalyst	Products	References
1-Vinylnaphthalene + butadiene	Ni(acac)$_2$ + Et$_3$Al + (o-tolyl-O)$_3$P	1-Naphthyldeca-1,4,8-triene	44, 59
p-Methoxystyrene + butadiene	Ni(acac)$_2$ + Et$_3$Al + (o-tolyl-O)$_3$P	1-p-Methoxyphenyldeca-1,4,8-triene	44
Chloroprene + ethylene	NiCl$_2$(Bu$_3$P)$_2$ + i-Bu$_2$AlCl	4-Chlorohexa-1,4-diene	41
	RhCl$_3$	2- or 3-Chlorohexa-2,4-diene	2
2,3-Dichlorobutadiene + ethylene	NiCl$_2$(Bu$_3$P)$_2$ + i-Bu$_2$AlCl	4,5-Dichlorohexa-1,4-diene	41
Methyl acrylate	RhCl$_3$ or RuCl$_3$	Dimethyl hex-2-enedioate	2
Methyl acrylate + ethylene	RuCl$_3$ or RhCl$_3$	Methyl pent-3-enoate 47%, C$_7$ acids 12%, C$_9$ acids 9%	2
Ethyl acrylate + butadiene	Ni(acac)$_2$ + Et$_3$Al + (o-tolyl-O)$_3$P	Ethyl hepta-4,6-dienoate	44
Ethyl acrylate + isoprene	Ni(acac)$_2$ + Et$_3$Al + (o-tolyl-O)$_3$P	Ethyl 6-methylhepta-4,6-dienoate	44
	Co(acac)$_2$ + Et$_3$Al	Ethyl 6-methylhepta-4,6-dienoate	126, 128
Ethyl acrylate + 2,3-dimethylbutadiene	Ni(acac)$_2$ + Et$_3$Al + (o-tolyl-O)$_3$P	Ethyl 5,6-dimethylhepta-4,6-dienoate	44

In view of the presence of the 1, 3-diene structure in these dimers it is not surprising that trimers, tetramers and higher oligomers are also formed. Catalyst systems with an aluminium to cobalt ratio higher than 3:1 convert butadiene solely to polymers[74]. Addition of potential ligands such as tri-(o-tolyl)phosphite diverts the dimerisation reaction to the production of cycloocta-1, 5-diene and 4-vinylcyclohexene accompanied by only traces of acyclic products[59]. Tri-π-allylcobalt also dimerises butadiene to 3-methylhepta-1, 4, 6-triene[94]. A complex of structure (32) has been isolated during work on the polymerisation of butadiene with a cobaltous chloride sodium borohydride catalytic system[61]. The same complex may be formed as an intermediate in the foregoing dimerisations of butadiene.

(32)

Bis(allyl)cobalt halides catalyse the polymerisation of butadiene to 1, 4-cis-polybutadiene[118]. The cobalt catalysts also effect "dienylation" reactions[128]. As noted with self-coupling of butadiene the products have the 1, 3-diene system as a result of hydrogen migration.

In contrast to the iron and cobalt catalysts linear butadiene dimers are obtained with nickel catalysts, (Table 2. 7). The nickel(0) catalysts which effect cyclodimerisation, yield linear dimers in the presence of alcohols[59, 82]. The best catalysts are the nickel carbonyl phosphite and phosphine complexes and especially bis(phosphite) acrylonitrilenickel(0) complexes. The principal products are octa-1, trans 3, 7-triene, octa-1, cis 3, trans 6-triene, and all trans-octa-2, 4, 6-triene. The use of deuterated hydroxylic solvents results in the formation of monodeuterated octa-1, 3, 7-triene[32]. Small amounts of 3-methylhepta-1, 4, 6-triene and cycloocta-1, 5-diene are formed together with higher oligomers in amounts depending on the reaction conditions. Of particular interest is the isolation of a new cyclic dimer of butadiene, namely 1-vinyl-3-methylenecyclopentane (33). The predominant dimer depends upon the complexing ligand of

(33)

the catalyst. For example, the 1, 3, 6-, 1, 3, 7- or 2, 4, 6-isomer of octatriene predominates when the ligand is triethylphosphite, tributylphosphine or phosphorus trimorpholide respectively.

Cooligomerisation of 1, 3-dienes with other olefins can also be effected with nickel(0) catalysts[41, 59]. In this case catalysts containing phosphorus ligands are said to be less satisfactory than nickel carbonyl or nickel(0) olefin complexes[59]. In some systems the products are derived from one molecule of olefin and two of butadiene, e.g.

Smaller amounts of other double-bond isomers are also formed. The use of less reactive olefins such as α-methylstyrene results in low yields of cooligomers as much of the butadiene is oligomerised.

Undoubtedly the intermediates involved in the case of linear dimerisation reactions are analogous to those detected in the cyclo-oligomerisation reactions. The derivation of the observed products from complexes such as (24) and (26) is readily envisaged. Particularly interesting is the contrast in catalytic behaviour between the bis(π-allyl) derivatives of nickel and palladium. While the former converts butadiene into cyclododecatriene, the palladium complex produces n-dodecatetraenes at a similar rate[118]. A possible, but not the only, explanation is that in the palladium equivalent of (24) the larger steric size of the metal atom precludes the linkage of the ends of the C_{12} chain.

π-Allylnickel complexes are also implicated in the coupling of allyl halides by reaction with nickel carbonyl in ether solution. As the published examples (Table 2. 8) demonstrate the products are principally derived from coupling at the least sterically encumbered carbon atoms, e.g.

While allyl halides and acetates undergo coupling the alcohols and phenyl ethers are inert. Homoallylic systems as typified by *endo*-norbornenyl acetate and but-3-enyl acetate are unreactive as are

TABLE 2.8 Coupling of Allylic Compounds

Allylic Compound	Products	References
Allyl acetate	Hexa-1,5-diene	6
β-Methallyl chloride	2,5-Dimethylhexa-1,5-diene	31
1-Chlorobut-2-ene } 3-Chlorobut-1-ene }	Octa-2,6-diene (78%) and 3-methylhepta-1, 5-diene (22%)	109, 110
1-Chloro-3-methylbut-2-ene } 3-Chloro-3-methylbut-1-ene }	2,7-Dimethylocta-2,6-diene (65%) and 3,3,6- trimethylhepta-1,5-diene (35%)	109, 110
1-Chloro-5-methoxypent-2-ene } 3-Chloro-5-methoxypent-2-ene }	1,10-Dimethoxydeca-3,7-diene	109, 110
3,6-Dichloro-5-ethoxyhex-1-ene } 1,6-Dichloro-5-ethoxyhex-2-ene }	1,12-Dichloro-2,11-diethoxydodeca-4,8-diene	95, 109
Methyl 4-bromocrotonate	Dimethyl octa-2,6-dienedioate	20
1-Chloro-1-cyanoprop-2-ene } 1-Chloro-3-cyanoprop-2-ene } 1-Bromo-3-cyanoprop-2-ene }	1,6-Dicyanohexa-1,5-diene	20
1-Chloro-4-cyanobut-2-ene	1,8-Dicyano-octa-2,6-diene (92%)	68
Cinnamyl acetate	1,6-Diphenylhexa-1,5-diene	6

57

also benzhydryl, phenacyl and benzoin acetates[6]. These observations are consistent with π-allyl ligand formation. Study of the coupling of ^{14}C allyl acetate showed that only partially rearranged hexa-1,5-diene was formed[26]. This is consistent with formation of a π-allyl-nickel complex which subsequently reacts with unrearranged allyl acetate.

If the reaction of allylbromide with nickel carbonyl is carried out in the presence of acrylonitrile then a mixture of hex-5-enonitrile, cis- and trans-hexa-2,5-dienonitrile, and hexa-2,4-dienonitrile are produced. The reaction of π-allylnickelbromide with acrylonitrile yields nickelbis(hex-5-enonitrile), which with water gives hex-5-enonitrile and on thermal decomposition the preceding product mixture. Methyl acrylate reacts analogously[19,27].

(34)

An interesting development of the allylic coupling reaction is its extension to intramolecular coupling of suitable bis-allylic systems. In a preliminary investigation, prior to an attempt to synthesise humulene (34), the internal coupling of the allylic bromide (35) gave the cyclododecatriene derivative (36)[24]. Either all trans or a 4:1 mixture of the all trans and cis,trans,trans-isomers of (35) gave a 2:1 mixture of (36) and its cis,trans,trans isomer.

$$BrCH_2.CH = \overset{\overset{\displaystyle CH_3}{|}}{C}.CH_2.CH_2.CH = CH.CH_2.CH_2.\overset{\overset{\displaystyle CH_3}{|}}{C} = CH.CH_2Br$$

(35)

(36)

(37)

π-Allylnickel halides themselves have considerable catalytic activity, especially in conjunction with an aluminium trihalide[118]. The activation by aluminium halides appears to arise from conversion of the dimeric π-allyl-nickel halides (37) into monomeric π-allylnickel tetrahaloaluminates. These catalysts convert ethylene into but-2-ene and propylene[94,118] into a mixture of 2-methylpentenes (80%) and hexenes (20%) simply by passing the olefin into a strongly cooled

solution of the catalyst in chlorobenzene. Addition of triphenylphosphine to give a nickel: phosphorus: aluminium molar ratio of 1:1:2
leads to the same products as above but lowers the amount of higher
oligomers formed. The introduction of strongly basic phosphines
like tri-isopropylphosphine results in propylene being converted into
a mixture of 2, 3-dimethylbutenes and 2-methylpentenes. Double
bond migration occurs under the dimerisation conditions. The importance of initial olefin coordination in the reaction is evidenced by the
blocking effect observed on the addition of further phosphine or an
unreactive olefin such as cycloocta-1, 5-diene.

Although the reactions of mono-olefins with titanium containing
Ziegler catalysts normally lead to polymer formation, interaction
with the methyltitanium trichloride-methylaluminium dichloride system at−70°C leads to predominant dimer and trimer formation[9a].
Typically, pent-1-ene yields 67% dimer and 18% trimers. The dimer
fraction comprises 90% 2-n-propylhept-1-ene and 8% of dec-3 and
4-enes, so that head-to-tail coupling is the major process. In the
case of but-1-ene the trimer fraction consists of 70% of 2, 4-diethy-
oct-1-ene and 30% of 6-ethyldec-2 and 3-enes. The major trimer
thus results from head to tail coupling and the others from a head
to tail followed by head to head linking.

Oligomerisation is also accompanied by some isomerisation of the
parent terminal olefin to the cis-2-isomer.

Only the methyl groups bonded to titanium are involved in the oligomerisation and are found in the high polymer fraction. It seems that
the catalyst is really $[CH_3Ti\,Cl_2]^+\;[CH_3Al\,Cl_3]^-$ with only the titanium cation being directly involved in the catalytic reaction although
other anions are less effective. The ionic nature of the catalytic
species is indicated by the relative effectiveness of different solvents which is chlorinated hydrocarbons > aromatic hydrocarbons >

paraffin hydrocarbons. Addition of olefin to the catalytic system causes a marked colour change from pale yellow-red to dark red suggesting immediate coordination of the olefin to titanium. The following stages have been proposed for this reaction involving chain growth and chain transfer steps.

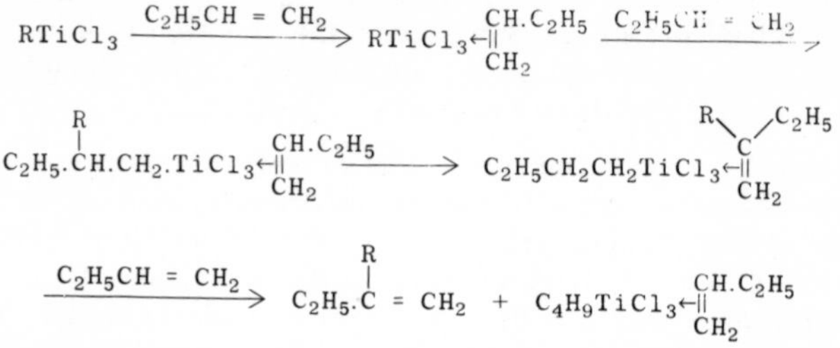

$$RTiCl_3 \xrightarrow{C_2H_5CH \ = \ CH_2} RTiCl_3 \!\! \leftarrow\!\! \underset{CH_2}{\overset{CH.C_2H_5}{\|}} \xrightarrow{C_2H_5CH \ = \ CH_2}$$

$$\underset{C_2H_5.CH.CH_2.TiCl_3}{\overset{R}{|}} \!\! \leftarrow\!\! \underset{CH_2}{\overset{CH.C_2H_5}{\|}} \longrightarrow C_2H_5CH_2CH_2TiCl_3 \!\! \leftarrow\!\! \underset{CH_2}{\overset{\overset{R}{\diagdown}C\overset{C_2H_5}{\diagup}}{\|}}$$

$$\xrightarrow{C_2H_5CH \ = \ CH_2} \underset{C_2H_5.C}{\overset{R}{|}} = CH_2 \ + \ C_4H_9TiCl_3 \!\! \leftarrow\!\! \underset{CH_2}{\overset{CH.C_2H_5}{\|}}$$

The coupling of olefins can also be effected in hydroxylic media using either rhodium trichloride or ruthenium chloride[1, 2, 107]. Ruthenium bromide and platinic chloride are much less efficient. It may be noted in passing that the ethylene palladium chloride complex catalyses the dimerisation of ethylene in non-hydroxylic solvents[106]. Results obtained in the dimerisation of ethylene indicate that but-1-ene is the initial product which under longer reaction times or higher temperatures is isomerised to the but-2-enes[2]. Similar double-bond migrations occur in the coupling of other olefins, (Table 2.7). Butadiene is converted into octa-2, 4, 6-triene probably by isomerisation of a precursor octa-1, 3, 6-triene. The predominance of the linear dimer in this reaction is noteworthy as is the formation of a linear dimer, dimethyl hex-2-enedioate, from methyl acrylate. Substitution at the double-bond of a mono-olefin appears to lower its reactivity so that propylene is much less reactive than ethylene. In the case of the 1, 3-dienes the reactivity depends on the position of substitution so that in order of decreasing

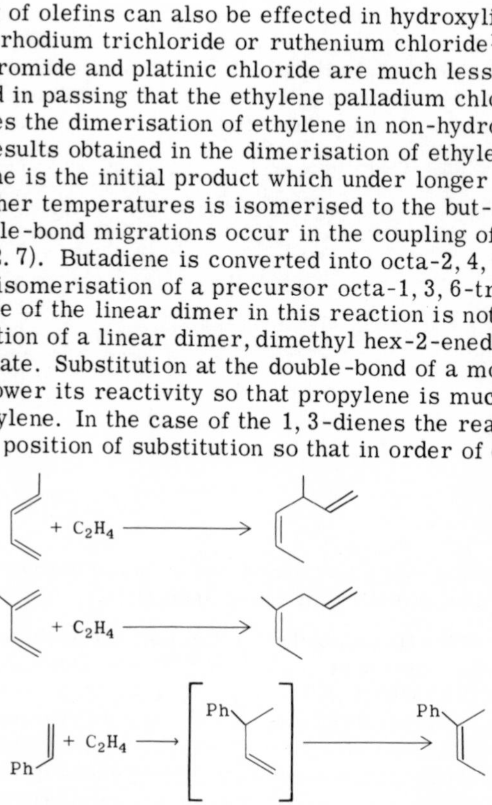

FIGURE 2.5

reactivity penta-1, 3-diene > butadiene > isoprene. In co-dimerisation reactions ethylene adds to the more highly substituted double-bond of 1, 3-dienes and the more substituted end of an olefinic double-bond, (Figure 2. 5).

The mechanism of the dimerisation of ethylene has been elucidated by a combination of kinetic and spectroscopic techniques[25]. Some of the intermediates have been isolated. The overall scheme is shown in Figure 2. 6. Direct reaction of the so-called rhodium (III) trichloride trihydrate with ethylene produces the bisethylene complex (39) of rhodium (I), which is also obtained by reaction of complexes such as (38) with hydrogen chloride. The complex (39) reacts rapidly with hydrogen chloride forming the ethylrhodium (III) compound (40). While complex (40) is stable at high pressures of ethylene and in

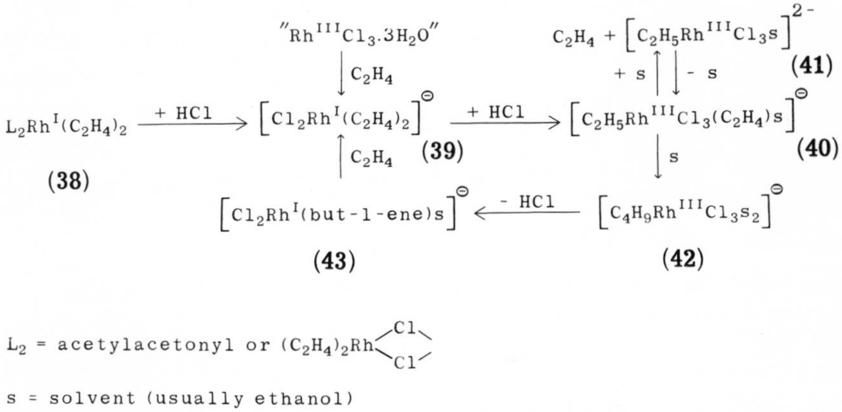

$$''Rh^{III}Cl_3.3H_2O''$$

$$\downarrow C_2H_4$$

$$C_2H_4 + \left[C_2H_5Rh^{III}Cl_3s\right]^{2-} \quad (41)$$

$$+ s \uparrow \downarrow - s$$

$$L_2Rh^I(C_2H_4)_2 \xrightarrow{\; + \; HCl \;} \left[Cl_2Rh^I(C_2H_4)_2\right]^{\ominus} \xrightarrow{\; + \; HCl \;} \left[C_2H_5Rh^{III}Cl_3(C_2H_4)s\right]^{\ominus}$$

$$(38) \qquad \uparrow C_2H_4 \quad (39) \qquad \qquad \downarrow s \qquad (40)$$

$$\left[Cl_2Rh^I(but-1-ene)s\right]^{\ominus} \xleftarrow{\; - \; HCl \;} \left[C_4H_9Rh^{III}Cl_3s_2\right]^{\ominus}$$

$$(43) \qquad\qquad\qquad (42)$$

$$L_2 = \text{acetylacetonyl or } (C_2H_4)_2Rh\underset{Cl}{\overset{Cl}{\diagup\negmedspace\diagdown}}$$

s = solvent (usually ethanol)

FIGURE 2. 6

solutions above 0. 1M in hydrogen chloride, at lower pressures of ethylene extensive dissociation to (41) occurs rapidly and extensively. Complex (40) rearranges by a slow, rate determining, chain growth reaction into the n-butyl rhodium (III) species (42). Rapid loss of hydrogen chloride from (42) produces the but-1-ene complex (43) of rhodium (I). Finally ethylene rapidly displaces the coordinated but-1-ene and solvent in complex (43) reforming (39).

Transition Metal Intermediates

REFERENCES

1. Alderson, T., U.S.P. 3, 013, 066; Chem. Abs., 1962, 57, 11016.

2. Alderson, T., Jenner, E. L. and Lindsey, R. V., J. Amer. Chem. Soc., 1965, 87, 5638.

3. Arnold, D. R., Trecker, D. J. and Whipple, E. B., J. Amer. Chem. Soc., 1965, 87, 2596.

4. Austin, A. A., Fr.P. 1, 377, 109; Chem. Abs., 1965, 62, 7657.

5. Badische Anilin-und Soda-Fabrik A. G., B.P. 944, 440; Chem. Abs., 1964, 60, 13164.

6. Bauld, N. L., Tetrahedron Letters, 1962, 859.

7. Benson, R. E., U.S.P. 2, 943, 116; Chem. Abs., 1961, 55, 2522.

8. Benson, R. E. and Lindsey, R. V., J. Amer. Chem. Soc., 1959, 81, 4247.

9. Benson, R. E. and Lindsey, R. V., J. Amer. Chem. Soc., 1959, 81, 4250

9a. Bestian, H. and Clauss, K., Angew. Chem. Internat. Edn., 1963, 2, 704.

10. Bieber, T. I., Chem. and Ind., 1957, 1126.

11. Bird, C. W., Colinese, D. L., Cookson, R. C., Hudec, J. and Williams, R. O., Tetrahedron Letters, 1961, 372.

12. Bird, C. W., Cookson, R. C. and Hudec, J., Chem. and Ind., 1960, 20.

13. Bosmajian, G., U.S.P. 3, 004, 081; Chem. Abs., 1962, 56, 4640.

14. Bosmajian, G., Burks, R. E. and Feazel, C. E., Ind and Eng. Chem. (Product Res. and Development), 1964, 3, 117.

15. Breil, H., Heimbach, P., Kröner, M., Müller, H. and Wilke, G., Makromol. Chem., 1963, 69, 18.

16. Burks, R. E. and Sekul, A. A., U.S.P. 2, 972, 640; Chem. Abs., 1962, 56, 4640.

17. Cairns, T. L., Engelhardt, V. A., Jackson, H. L., Kalb, G. H. and Sauer, J. C., J. Amer. Chem. Soc., 1952, 74, 5636.

17a. Cannell, L. G., Tetrahedron Letters, 1966, 5967

18. Carbonaro, A., Bonfardeci, A. and Porri, L., Fr.P. 1, 393, 071; Chem. Abs., 1965, 63, 1715.

19. Chiusoli, G. P., Chimica e Industria, 1961, 43, 365

20. Chiusoli, G. P. and Cometti, G., Chimica e Industria, 1963, 45, 401.

21. Cities Service Research and Development Co., B.P. 944, 574; Chem. Abs., 1964, 60, 10569.

22. Cities Service Research and Development Co., B.P. 947, 656; Chem. Abs., 1964, 60, 14406.

23. Consortium für Elektrochemische Industrie G.m.b.H., B.P. 887, 362; Chem. Abs., 1963, 58, 3521.

24. Corey, E. J. and Hamanaka, E., J. Amer. Chem. Soc., 1964, 86, 1641.

25. Cramer, R., J. Amer. Chem. Soc., 1965, 87, 4717.

26. Dubini, M., Chiusoli, G. P. and Montino, F., Tetrahedron Letters, 1963, 1591.

27. Dubini, M., Montino, F. and Chiusoli, G. P., Chimica e Industria, 1965, 47, 839.

28. E. I. Du Pont de Nemours and Co., B.P. 812, 901; Chem. Abs., 1960, 54, 7669.

29. E. I. Du Pont de Nemours and Co., B.P. 812, 902; Chem. Abs., 1960, 54, 7591.

30. Esso Research and Engineering Co., B.P. 903, 651; Chem. Abs., 1963, 58, 459.

31. I. G. Farben A. G., Belg. P. 448, 884; Chem. Abs., 1947, 41, 6576.

32. Feldman, J., Frampton, O., Saffer, B. and Thomas, M., Amer. Chem. Soc., Div. Petrol Chem., Preprints, 1964, 9, A55; Chem. Abs., 1966, 64, 14076.

33. Feldman, J. and Saffer, B.A., U.S.P. 3, 227, 767; Chem. Abs., 1966, 64, 9614.

34. Foster, R. E., U.S.P. 2, 504, 016; Chem. Abs., 1950, 44, 7873.

35. Friedrich, H. H., Schweckendieck, W. and Sepp, K., G.P. 1, 005, 954; Chem. Abs., 1959, 53, 17906.

36. Ghymes, G., Gröbler, A., Simon, A., Kada, I. and Andor, I., Magyar Kém. Lapja, 1965, 20, 570.

37. Hata, G., Chem. and Ind., 1965, 223.

38. Hidai, M., Uchida, Y. and Misono, A., Bull. Chem. Soc. Japan, 1965, 38, 1243.

39. Jolly, P. W., Stone, F.G.A. and Mackenzie, K., J. Chem. Soc., 1965, 6416.

40. Kasatkina, N. G. and Piastro, V. D., Vestnik Leningrad Univ., Ser. Fiz. i Khim., 1960, **15**, 140.

41. Kealy, T. J., Fr.P. 1, 388, 305; Chem. Abs., 1965, **63**, 6858.

42. Kutepow, N., Seibt, H. and Meier, F., G.P. 1, 144, 268; Chem. Abs., 1963, **59**, 3790.

43. Lapporte, S. J., Fr.P. 1, 398, 383; Chem. Abs., 1965, **63**, 14726.

44. Lautenschlager, H., Scharf, E., Wittenberg, D. and Müller, H., Belg.P. 622, 195; Chem. Abs., 1963, **59**, 9879.

45. Lemal, D. M. and Shim, K. S., Tetrahedron Letters, 1961, 368.

46. Leto, J. R. and Fiene, M. L., J. Amer. Chem. Soc., 1961, **83**, 2944.

47. Leto, J. R. and Fiene, M. L., U.S.P. 3, 076, 016; Chem. Abs., 1963, **59**, 6276.

48. Levine, R., Fr.P. 1, 321, 454; Chem. Abs., 1963, **59**, 11293.

49. Longiave, C., Castelli, R. and Andreetta, A., Belg.P. 615, 375; Chem. Abs., 1963, **58**, 2382.

50. Martin, H. and Vohwinkel, F., Chem. Ber., 1961, **94**, 2416.

51. Misono, A., Uchida, Y. and Saito, T., Jap.P. 21, 410('65); Chem. Abs, 1966, **64**, 1954.

52. Müller, H., G.P. 1, 080, 547; Chem. Abs., 1961, **55**, 16452.

53. Müller, H., G.P. 1, 080, 548; Chem. Abs., 1961, **55**, 14986.

54. Müller, H., G.P. 1, 095, 819; Chem. Abs., 1962, **56**, 8592.

55. Müller, H., G.P. 1, 097, 982; Chem. Abs., 1961, **55**, 25808.

56. Müller, H., G.P. 1, 106, 758; Chem. Abs., 1962, **56**, 2352.

57. Müller, H., G.P. 1, 126, 864; Chem. Abs., 1962, **57**, 8461.

58. Müller, H., Scharf, D. and Wittenberg, D., Fr.P. 1, 379, 251; Chem. Abs., 1965, **62**, 9007.

59. Müller, H., Wittenberg, D., Seibt, H. and Scharf, E., Angew. Chem., Internat. Edn., 1965, **4**, 327.

60. Müller, H., Wittenberg, D. and Scharf, E., G.P. 1, 204, 669; Chem. Abs., 1966, **64**, 3380.

61. Natta, G., Giannini, U., Pino, P. and Cassata, A., Chimica e Industria, 1965, **47**, 524.

62. Neutzel, K., G.P. 1, 165, 590; Chem. Abs., 1964, **60**, 15752.

63. Nicolson, A. and Turner, A. H., B.P. 1, 007, 209; Chem. Abs. 1966, **64**, 1984.

64. Otsuka, S. and Taketomi, K., European Polymer J., 1966, **2**, 289.

65. Otsuka, S., Taketomi, T., Imaizumi, F., Kikuchi, T., Nagaoka, I. and Muranishi, S., Jap.P. 23, 326('65); Chem. Abs., 1966, **64**, 4967.

66. Otsuka, S., Taketomi, T., and Kikuchi, T., J. Amer. Chem. Soc., 1963, **85**, 3709; J. Chem. Soc. Japan, Ind. Chem. Sect., 1963, **66**, 1094.

67. Pisman, I. I., Chernikova, I. M. and Dalin, M. A., U.S.S.R. 174, 620; Chem. Abs., 1966, **64**, 1954.

68. Prichard, W. W. and Whitman, G. M., U.S.P. 2, 524, 833; Chem. Abs., 1951, **45**, 1618.

69. Pruett, R. L., U.S.P. 3, 168, 581; Chem. Abs., 1965, **62**, 11704.

70. Reed, H. W. B., J. Chem. Soc., 1954, 1931.

71. Reed, H. W. B., B.P. 701, 106; Chem. Abs., 1955, **49**, 2495.

72. Reed, H. W. B., U.S.P. 2, 686, 209; Chem. Abs., 1955, **49**, 15957.

73. Reppe, W. and Schweckendiek, W. J., Annalen, 1948, **560**, 104.

74. Saito, T., Ono, T., Uchida, Y. and Misono, A., J. Chem. Soc. Japan, Ind. Chem. Sect., 1963, **66**, 1099; Bull. Chem. Soc. Japan, 1964, **37**, 105.

75. Saito, T., Uchida, Y. and Misono, A., Bull. Chem. Soc. Japan, 1965, **38**, 1397.

76. Sauer, J. C. and Cairns, T. L., J. Amer. Chem. Soc., 1957, **79**, 2659.

77. Schneider, K. and Schnell, H., G.P. 1, 086, 226; Chem. Abs., 1961, **55**, 27150.

78. Schrauzer, G. N., J. Amer. Chem. Soc., 1959, **81**, 5310.

78a. Schrauzer, G. N., Bastian, B. N. and Fosselius, G. A., J. Amer. Chem. Soc., 1966, **88**, 4890.

79. Schrauzer, G. N. and Eichler, S., Chem. Ber., 1962, **95**, 2764.

80. Schrauzer, G. N. and Eichler, S., G.P. 1, 186, 052; Chem. Abs., 1965, **62**, 16086.

81. Schrauzer, G. N. and Glockner, P., Chem. Ber., 1964, **97**, 2451.

82. Seibt, H. and Kutepow, N., Belg.P. 635, 483; Chem. Abs., 1964, **61**, 11891.

83. Sekul, A. A. and Sellers, H. G., U.S.P. 2, 964, 575; Chem. Abs., 1961, **55**, 14333.

84. Sellers, H. G. and Sekul, A., U.S.P. 2, 991, 317; Chem. Abs., 1962, 56, 3374.

85. Shell Internationale Research Maatschappij N. V., Belg. P. 626, 407; Chem. Abs. 1964, 60, 13164.

86. Shell Internationale Research Maatschappij N.V., Neth. Appl., 294, 637; Chem. Abs., 1965, 63, 14698.

87. Shell Internationale Research Maatschappij. N. V., Neth. Appl., 6, 413, 303; Chem. Abs., 1965, 63, 17934.

88. Shell Internationale Research Maatschappij. N. V., Neth. Appl., 6, 506, 276; Chem. Abs., 1966, 64, 11104.

89. Shikata, K., Miura, Y., Nakao, S. and Azuma, K., J. Chem. Soc. Japan, Ind. Chem. Sect., 1965, 68, 2266.

90. Stamicarbon N. V., B.P. 856, 858; Chem. Abs., 1961, 55, 13913.

91. Studiengesellschaft Kohle m.b.H., B.P. 860, 377; Chem. Abs., 1961, 55, 17546.

92. Studiengesellschaft Kohle m.b.H., B.P. 917, 013; Chem. Abs., 1963, 58, 13815.

93. Studiengesellschaft Kohle m.b.H., Belg.P. 619, 490; Chem. Abs., 1963, 59, 11293.

94. Studiengesellschaft Kohle m.b.H., Neth. Appl. 6, 409, 179; Chem. Abs., 1965, 63, 5770.

95. Tabet, G. E., U.S.P. 2, 570, 887; Chem. Abs., 1952, 46, 3071.

96. Takahashi, H., Tai, S. and Yamaguchi, M., J. Org. Chem., 1965, 30, 1661.

97. Takahashi, H. and Yamaguchi, M., J. Org. Chem., 1963, 28, 1409.

98. Takahashi, H. and Yamaguchi, M., Bull. Osaka Ind. Res. Inst., 1964, 15, 271.

99. Tanaka, S., Mabuchi, K. and Sato, T., Jap.P. 9003('65); Chem. Abs., 1965, 63, 6858.

100. Tani, K. and Yuguchi, S., Jap.P. 21, 618('65); Chem. Abs., 1966, 64, 1984.

101. Tani, K. and Yuguchi, S., Jap.P. 21, 619('65); Chem. Abs., 1966, 64, 1984.

102. Tepenitsyna, E. P., Dorogova, N. K., and Farberov, M. I., Neftekhimiya, 1962, 2, 604.

103. Toyo Rayon Co. Ltd., Fr.P. 1, 386, 994; Chem. Abs., 1965, 62, 13063.

Oligomerisation of Olefins

104. Toyo Rayon Co. Ltd., Fr.P. 1, 417, 455; Chem. Abs., 1966, 64, 6491.

105. Treboganov, A. D., Mitsner, B. I., Zinkevich, E. P., Kraerskii, A. A. and Preobrazhenskii, N. A., Zhur. org. Khim., 1965, 1, 1583.

106. Van Gemert, J. T. and Wilkinson, P. R., J. Phys. Chem., 1964, 68, 645.

107. Verbanc, J. I., Fr.P. 1, 319, 578; Chem. Abs., 1963, 59, 9785.

108. Vogel, E., Annalen, 1958, 615, 1.

109. Webb, I. D., U.S.P. 2, 654, 787; Chem. Abs., 1954, 48, 10048.

110. Webb, I. D. and Borcherdt, G. T. J. Amer. Chem. Soc., 1951, 73, 2654.

111. Weber, H., G.P. 1, 112, 069; Chem. Abs., 1962, 56, 8593.

112. Weber, H., Ring, W., Hochmuth, U. and Franke, W., Annalen, 1965, 681, 10.

113. Wilke, G., J. Polymer. Sci., 1959, 38, 45.

114. Wilke, G., G.P. 1, 078, 108; B.P. 837, 267; Chem. Abs., 1960, 54, 24391.

115. Wilke, G., G.P. 1, 050, 333; 1, 056, 123; Chem. Abs., 1961, 55, 4393; 14333.

116. Wilke, G., Angew. Chem., Internat. Edn., 1963, 2, 105.

117. Wilke, G. and Bogdanovic, B., Angew. Chem., 1961, 73, 756.

118. Wilke, G., Bogdanovic, B., Hardt, P., Heimbach, P., Keim, W., Kröner, M., Oberkirch, W., Tanaka, K., Steinrücke, E., Walter, D. and Zimmermann, H., Angew. Chem., Internat. Edn., 1966, 5, 151.

119. Wilke, G. and Heimbach, P., Belg.P. 630, 046; Chem. Abs., 1964, 60, 15752.

120. Wilke, G. and Herrmann, G., Angew. Chem., Internat. Edn., 1962, 1, 549.

121. Wilke, G. and Kröner, M., Angew. Chem., 1959, 71, 574.

122. Wilke, G., Kröner, M. and Bogdanovic, B., Angew. Chem., 1961, 73, 755.

123. Wilke, G. and Müller, H., G.P. 1, 043, 329; Chem. Abs., 1961, 55, 2523.

124. Wilke, G. and Müller, E. W., G.P. 1, 140, 569; Chem. Abs. 1963, 58, 11214.

125. Wilke, G., Müller, E. W., Kröner, M., Heimbach, P. and Breil, H., Fr.P. 1, 320, 729; Chem. Abs., 1963, **59**, 14026.

126. Wittenberg, D., Belg.P. 618, 625; Chem. Abs., 1963, **59**, 2644.

127. Wittenberg, D., G.P. 1, 133, 368; Chem. Abs., 1963, **58**, 3332.

128. Wittenberg, D., Angew. Chem., Internat. Edn., 1964, **3**, 153.

129. Wittenberg, D., Lautenschlager, H., Kutepow, N., Meier, F. and Seibt, H., Belg.P. 621, 730; Fr.P. 1, 350, 644; Chem. Abs., 1963, **59**, 11292.

130. Wittenberg, D. and Müller, H., G.P. 1, 109, 674; Chem. Abs., 1962, **56**, 7178.

131. Wittenberg, D. and Müller, H., G.P. 1, 137, 007; Chem. Abs., 1963, **58**, 7846.

132. Yamamoto, A., Morifuji, K., Ikeda, S., Saito, T., Uchida, Y. and Misono, A., J. Amer. Chem. Soc., 1965, **87**, 4652.

133. Yoshida, T., Iwamoto, M. and Yuguchi, S., Jap.P. 12, 697('65); Chem. Abs., 1966, **64**, 11103.

134. Yoshida, T. and Yuguchi, S., Jap.P. 21, 617('65); Chem. Abs., 1966, **64**, 1983.

135. Zakharkin, L. I., Doklady Akad. Nauk S.S.S.R., 1960, **131**, 1069.

136. Zakharkin, L. I. and Korneva, V. V., Chem. Abs., 1965, **62**, 6389.

137. Zakharkin, L. I. and Zhigareva, G. G., Izvest. Akad. Nauk, S.S.S.R., Ser. khim., 1963, 386.

138. Zakharkin, L. I. and Zhigareva, G. G., Izvest. Akad. Nauk, S.S.S.R., Ser. khim., 1964, 168.

139. Ziegler, K., Sauer, H., Bruns, L., Froitzheim-Kühlhorn, H., and Schneider, J., Annalen, 1954, **589**, 122.

140. Ziegler, K., and Wilms, H., Annalen, 1950, **567**, 1.

CHAPTER 3

HYDROGEN MIGRATION REACTIONS

Hydrogen migration reactions are sometimes encountered when working with transition metal catalysts. Frequently, as with double-bond migration in the hydroformylation reaction, they are regarded as a nuisance. However, the very mild conditions under which they can be effected has begun to attract attention to their synthetic applications.

From a mechanistic viewpoint the most extensively investigated catalysts are compounds of the platinum metals. The most efficient of these contain rhodium (I) or (III). A common feature of such catalysts is the need for a co-catalyst. This is usually ethanol, but secondary and tertiary alcohols, ethers, ketones and carboxylic acids can be used[27]. The rate of reaction decreases in the order ethanol > isopropanol > t-butanol ~ acetic acid. Hydrogen can also function as a co-catalyst. The use of deuterated ethanol results in the incorporation of deuterium into isomerised olefins, but not with acetic acid[5,14]. Rearrangement of a mixture of inactive hex-1-ene and oct-1-ene tritiated in the allylic position results in transfer of tritium from the octene to the rearranged hexenes[5].

A characteristic feature of most of these systems is the preferential formation of the cis-2-isomer from a terminal olefin in the early stages of reaction[27]. An analogous situation arises in the initial accumulation of cis-3-isomer during the isomerisation of the 2-isomer. The migration of the double bond from position 2 to 3 is slower than from 1 to 2. Eventually the isomer distribution approaches the thermodynamically predicted one. As the first stage in the isomerisation entails the coordination of the olefin to the metal atom, vide infra, some olefins are often difficult to isomerise even though they are not the most thermodynamically stable isomers. As anticipated the rate of isomerisation of an olefin such as 3, 3-dideutero-oct-1-ene is slower than that of oct-1-ene[16]. Conversely the rearrangement of a mixture of 1, 2-dideutero and 1, 1, 2-trideutero-hex-1-ene is about three times faster than for hex-1-ene.

Three different mechanisms have been considered for double bond isomerisation by transition metal compounds, all of which envisage the initial coordination of the olefin to the metal species but differ in the mechanism by which hydrogen migration occurs. One of these postulates hydrogen transfer from the allylic position of the olefin to the coordinated metal forming an intermediate π-allyl metal hydride (1). Transfer of the hydride from the metal to the other end of the π-allyl system generates the rearranged olefin complex (2). An alternative suggestion is that the initial olefin complex isomerises by way of a carbene complex (3)[15].

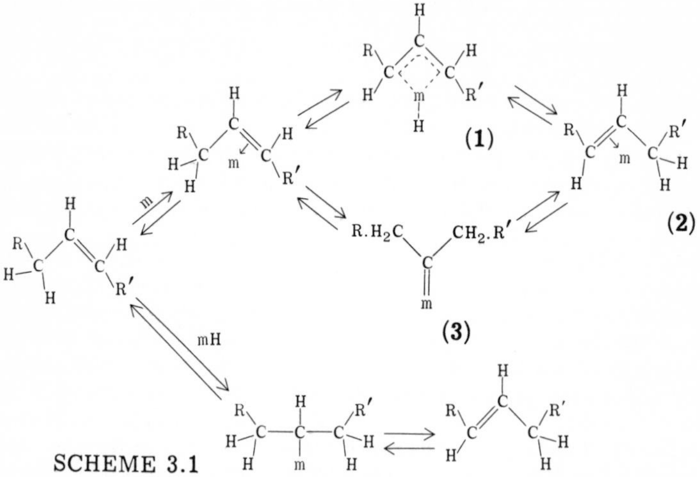

SCHEME 3.1

Although both of the preceding mechanisms have received some experimental support it seems that most of it can be reconciled with a variant of the metal hydride addition-elimination mechanism[13a], cf. Scheme 3.1. Strong support for a mechanism of this type has been obtained from a study of the isomerisation of butenes with a rhodium catalyst[13]. The catalyst is derived from rhodium trichloride, $[(C_2H_4)_2RhCl]_2$ or $[(C_2H_4)_2Rhacac]$. In the latter case addition of hydrogen chloride is essential for catalytic activity, other acids such as sulphuric are ineffective. In the absence of hydrogen or a co-catalyst, pretreatment with olefin generates the necessary hydride species. The rhodium-catalysed isomerisation of but-1-ene is retarded by other olefins in the order acrylonitrile > butadiene ≫ acetylene > vinyl chloride ~ ethylene > tetrafluoroethylene. A similar retarding effect is displayed by other compounds which coordinate strongly to rhodium. The relative rate constants for the isomerisation of the individual butenes at 25°C shows the reason for the initial accumulation of the cis-2-isomer.

The essential features of the mechanism proposed for the isomerisation is depicted in Scheme 3.2. Much of the detailed evidence on which the proposed sequence is based is derived from the study of deuterium transfer which occurs in deuterated solvents or accompanies the isomerisation of deuterated olefins. Not surprisingly there is a

70

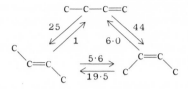

close relationship between this sequence and that deduced for the rhodium-catalysed coupling of olefins (cf. p. 60). The inclusion of stannous chloride results in a higher isomerisation to deuteration ratio but at tin to rhodium ratios above 6 to 1, the isomerisation is stopped[13a]. The tris-(diethylphenylphosphine)rhodium trichloride complex is reported to be a good catalyst for the isomerisation of cycloocta-1, 5-diene[40], and rhodium trichloride with the same diolefin yields the complex $[(C_8H_{12})RhCl]_2$ and converts all of the uncomplexed olefin into the 1, 3- and 1, 4-isomers. Although up to 35% of the cycloocta-1, 4-diene has been obtained other workers[41] were unable to detect this isomer. The same complex is formed from cycloocta-1, 3-diene[41], but it is catalytically inactive[40].

$$Rh^{I}L_3 + HCl \rightleftharpoons HRh^{III}L_3Cl \qquad (i)$$
$$\uparrow \qquad\qquad\qquad \uparrow$$
$$C_2H_4 \qquad\qquad\qquad C_2H_4$$

$$HRh^{III}L_3Cl + CH_2 = CH.C_2H_5 \rightleftharpoons HRh^{III}L_3Cl + C_2H_4$$
$$\uparrow \qquad\qquad\qquad\qquad\qquad \uparrow$$
$$C_2H_4 \qquad\qquad\qquad\qquad CH_2{=}CH{\cdot}C_2H_5 \qquad (ii)$$

$$HRh^{III}L_3Cl \rightleftharpoons Rh^{III}L_{3\,or\,4}Cl$$
$$\uparrow \qquad\qquad\qquad |$$
$$CH_2{=}CH{\cdot}C_2H_5 \qquad CH_3{\cdot}CH{\cdot}C_2H_5 \qquad (iii)$$

$$Rh^{III}L_{3\,or\,4}Cl \rightleftharpoons HRh^{III}L_3Cl$$
$$| \qquad\qquad\qquad \uparrow$$
$$CH_3.CH.CH_2.CH_3 \qquad CH_3CH{=}CH{\cdot}CH_3 \qquad (iv)$$

$$HRh^{III}L_3Cl \rightleftharpoons Rh^{I}L_3 + HCl$$
$$\uparrow \qquad\qquad\qquad \uparrow$$
$$CH_3CH{=}CH{\cdot}CH_3 \qquad CH_3CH{=}CH{\cdot}CH_3 \qquad (v)$$

SCHEME 3.2

Although most of the information on iridium catalysed double bond migration concerns phosphine complexes, the trichloride catalyses the isomerisation of hex-1-ene[27] and with cycloocta-1, 5-diene or the 1, 3-isomer forms the complex $[(C_8H_{12})IrHCl_2]$, which liberates the 1, 5-diene on treatment with aqueous potassium cyanide[40,41]. All of the uncomplexed octadiene is converted to the 1, 3-diene. The same complex $[(C_8H_{12})_2IrSnCl_3]$ is unexpectedly obtained from the reaction of sodium hexachloroiridate and stannous chloride with 4-vinylcyclohexene, cycloocta-1,3-diene or cycloocta-1,5-diene[49]. A different complex with the composition $[(C_7H_8)_2IrSnCl_3]$ is formed from either norbornadiene or cycloheptatriene.

Iridium complexes of the type $[R_3P)_3IrCl_3]$ where R is ethyl or butyl catalyse the isomerisation of pure oct-1-ene to oct-2-ene and a little or the 3 isomer at 160°-170°C[12]. The catalyst is recovered unchanged, but if unpurified oct-1-ene is employed rapid isomerisation occurs at 120°C and a good yield of $[(Et_3P)_3IrHCl_2]$ is obtained.[12] These observations indicate that a preformed iridium hydride assists olefin isomerisation but is not necessarily essential. Not surprisingly the hydride complex $[(Et_2PhP)_3IrHCl_2]$ also rapidly converts oct-1-ene in refluxing benzene solution to a mixture of isomers. It also catalyses the isomerisation of cycloocta-1,5-diene at 130°C[40]. After 22 minutes a mixture containing 30% of the 1,4 and 50% of the 1,3-isomer is obtained. A 90% conversion to the 1,3-isomer is obtained after two hours. This complex is about five times more efficient than $[(Et_2PhP)_3IrCl_3]$ and the catalytic activity of both complexes is curbed by free diethylphenylphosphine. The catalytic efficiency of these complexes decreases in the sequence $[(Et_2PhP)_3IrHCl_2] > [(Et_3P)_3IrCl_3] > [(Et_2PhP)_3IrCl_3] > [(Me_2PhP)_3IrCl_3]$ which is opposite to the probable order of increasing affinity of the phosphine ligands.

The complex $[(Ph_3P)_3IrH_3]$ is also an efficient catalyst but is inhibited by addition of triphenylphosphine[12]. Observations of this type together with the isolation of starting complexes suggest that the initial step in the isomerisation involves replacement of a phosphine ligand by the olefin. The high trans effect of the hydride ligand should facilitate the replacement. When the complex $[(Ph_3P)_3IrD_3]$ is used for isomerisation of oct-1-ene it is converted into $[(Ph_3P)_3IrH_3]$. The initial rate of isomerisation is much slower with the deutero complex than with the corresponding hydrido one. While it seems likely that the exchange of hydrogen for deuterium occurs during the isomerisation process the possibility that it results from a side-reaction is not readily excluded.

Hydrogen transfer from isopropanol to ketones is catalysed by chloroiridic acid and trimethyl phosphite[26]. The synthetic value of this procedure arises from the high yields of axial alcohols produced. Cholestanone, 3-t-butylcyclohexanone and 3,3,5-trimethylcyclohexanone give 92, 98 and 99% yields respectively of the corresponding axial alcohols. The replacement of trimethyl phosphite by dimethyl sulphoxide results in an axial to equatorial ratio of alcohols similar to that obtained by heterogeneous hydrogenation. Cholestanone gives 3α and 3β-cholestanols in a 69:31 ratio and 4-t-butylcyclohexanone a 78:22 mixture of the *cis* and *trans* alcohols. In all cases acetone is formed indicating that *iso*-propanol is the hydrogen source. Experiments using 2-deutero *iso*-propanol show that it is the C2 hydrogen of the *iso*-propanol which is transferred to the carbonyl carbon atom. It is perhaps apposite to note that reaction of potassium hexachloroiridate (IV) with triphenylphosphine yields the complex $[(Ph_3P)_3IrHCl_2]$ in which the hydride originates from the α-carbon atom of the alcohol[47].

Palladium-catalysed double bond migration has received much attention. Under non-hydroxylic conditions the activity of palladous chloride appears to vary with the sample[44]. This problem does not arise with its readily accessible bis-benzonitrile complex. The ethylene or cyclohexene complexes are equally effective but the complex *trans*-$[(Me_2PhP)_2PdI_2]$ is relatively poor[40]. In apparent contradiction to stepwise isomerisation by addition and elimination of hydride is the observation that during the isomerisation of 4-methylpent-1-ene, 2-methylpent-1-ene is formed at a rate faster than is consistent with 2-methylpent-2-ene being the sole precursor[44]. It is suggested[13a] that part of the 4-methylpent-1-ene is cleaved into two propylene units which recombine to give 2-methylpent-1-ene (cf. Chapter 2). Such a process would also account for the formation of linear hexenes which were observed in the reaction products. Another case of carbon-carbon bond cleavage occurs in the formation of $[(1,5\text{-cyclooctadiene})PdCl_2]$ from 4-vinylcyclohexene and palladous chloride[24].

The rate of isomerisation of but-1-ene is much slower in benzene, toluene, chlorobenzene or dichloroethane, but is faster in chloroform or nitromethane and is fastest in nitrobenzene[38]. The *trans* to *cis* ratio is solvent dependent being 1·8 in benzene, 2 to 2·8 in the preceding solvents, 3·2 in water and 4 to 6 in dimethylformamide.

Deactivation of palladium catalysts can arise by formation of π-allylpalladium complexes[28], which are most readily formed from branched olefins. In accord with this it has been observed that of the four 2-methylpentenes the one with the most branching at the olefinic double bond causes the most rapid catalyst deactivation[44].

The combination of a palladium compound with a carboxylic acid has also been employed. The best acid has been shown to be trifluoroacetic[13a]. There appears to be a marked difference between this system and the palladium catalysed ones previously discussed. The isomerisation of 1-olefins with palladous chloride complexes exhibits the usual preferential formation of *cis*- rather than *trans*-2-isomer in the early stages[44]. The relative rate constants for interconversion of butenes with a lithium tetrachloropalladate (II)-trifluoroacetic acid system show a different pattern of behaviour, to that observed with rhodium catalysts.

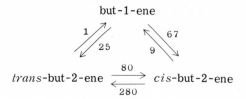

Only a small amount of proton exchange occurs between olefin and solvent under these conditions[13a,14]. The origin of the palladium hydride species appears[13a] to arise as suggested earlier[27], i.e.

$$\begin{array}{c}\backslash\diagup H\\ C\\ \|\!\!\longrightarrow Pd^{II}Cl\\ C\\ \diagup\backslash\end{array} \quad\longrightarrow\quad \begin{array}{c}\backslash\diagup Cl\\ C\\ \|\!\!\longrightarrow Pd^{II}\ \text{-}\ H\\ C\\ \diagup\backslash\end{array}$$

Isomerisation of cis-but-2-ene in the presence of tetradeuteroethylene with Li_2PdCl_4/CF_3CO_2H yields partially deuterated butenes and vinyl chloride[13a]. The palladium catalysts are very susceptible to inhibition by chloride and $SnCl_3^-$ anions.

A catalytic system comprising a methanolic solution of tetrakis(triethylphosphite)nickel(O) containing sulphuric acid very rapidly isomerises but-1-ene[13a]. The relative rate constants for the isomerisation of the butenes are closely similar to those found for the palladium system and very little hydrogen exchange occurs with the solvent.

A range of platinum compounds have also been found to catalyse double bond migration. These include chloroplatinic acid[21,22] $[(C_2H_4)Pt\ Cl_2]_2$[27] and $trans\ [(Et_3P)_2PtHCl]$[13a,40]. All, except the latter complex which needs rather drastic conditions, appear to need a cocatalyst and once again a metal hydride addition-elimination mechanism is implicated[13a]. Hydrogen only acts as a cocatalyst in the presence of stannous chloride. The presence of stannous chloride appears to result in a higher proportion of $trans$ isomer[9,13a] than is obtained in its absence[27]. The proportions of isomers obtained from isomerisation of butenes using the $H_2PtCl_6/SnCl_2/H_2$ system is similar to that obtained with the acidic palladium system although comparison of the respective sets of relative rate constants shows that a somewhat different route is involved[13a].

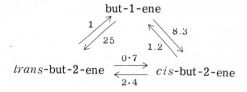

The exchange of hydrogen with solvent is much higher with this system than with rhodium catalysts.

Despite a contradictory report[27] ruthenium chloride also appears to catalyse double bond migration[1,10]. Both intermolecular and intramolecular hydrogen transfer reactions catalysed by ruthenium chloride have been observed with allyl alcohols[39].

$$2\ CH_2 = CH.CH_2OH \longrightarrow CH_3.CH = CH_2 + CH_2 = CH.CHO + H_2O$$

$$CH_2 = CH.CH_2OH \longrightarrow CH_3.CH_2.CHO$$

Ruthenium (III) chloride is rapidly reduced to ruthenium (II) which is apparently the catalytically active oxidation state. Rhodium (III) and palladium (II) also show some ability to catalyse reactions of this type. Other allylic alcohols undergo comparable reactions.

$$CH_2 = \overset{\overset{\displaystyle CH_3}{|}}{C}.CH_2OH \longrightarrow CH_2 = C(CH_3)_2 + CH_2 = \overset{\overset{\displaystyle CH_3}{|}}{C}.CHO$$

$$+ (CH_3)_2CH \quad CHO$$

$$CH_3CH = CH.CH_2OH \longrightarrow CH_3CH_2COCH_3 + n\text{-butenes}$$

$$HOCH_2.CH = CH.CH_2OH \longrightarrow CH_3CH = CH.CHO$$

$$HC \equiv C.CHO \longrightarrow C_2H_4 + CO$$

A comparison of a variety of metal carbonyls has demonstrated the particular efficacy of iron carbonyls either on heating or irradiation with ultraviolet light for the isomerisation of n-undec-1-ene[4]. Di-iron nonacarbonyl, the normal product of irradiation of iron pentacarbonyl, functions efficiently in the dark[6]. Comparison with other metal carbonyls shows that dicobalt octacarbonyl and di-osmium nonacarbonyl have a little activity but that nickel carbonyl, dimanganese decacarbonyl and the hexacarbonyls of chromium, molybdenum and tungsten are ineffective[2,3,6]. Iron carbonyls are best employed in an alkane solvent and the rate of isomerisation depends on the light intensity. Carbon monoxide has an inhibitory effect[6,37].

With diolefins the double-bonds usually migrate into conjugation and the resulting 1,3-diene is converted into a metal carbonyl complex so that in this case the catalyst is consumed[2,18,33]. The diene can be liberated from iron tricarbonyl complexes by oxidation with ferric chloride. A survey of the reactions of a series of diolefins with iron pentacarbonyl indicates that they are converted to the least sterically crowded complex, cf. Figure 3.1. Only cis, trans-hexa-2,4-diene produces an iron tricarbonyl complex with a substituent in an *anti* position. Substituents in the *anti* position of the planar diene ligand cause considerable steric strain in the complex. This results either from repulsion between the *anti* substituents themselves or between

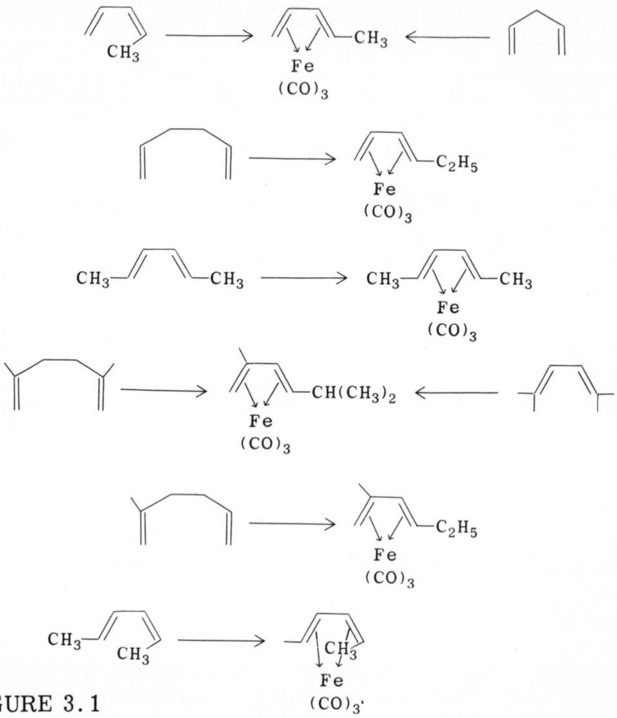

FIGURE 3.1

an *anti* substituent and the iron atom or attached carbonyl group. An interesting extension is the conversion of α-cyclopropylstyrenes into 2-arylpenta-1,3-dienes[43].

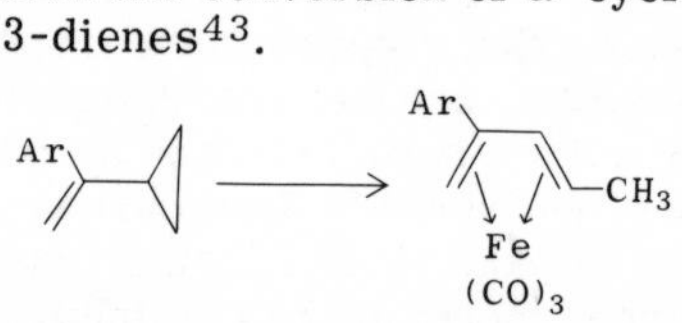

Another case of ring fission, only without hydrogen migration occurring, has been observed in the reaction of 8,9-dihydroindene with molybdenum hexacarbonyl yielding cyclononatetraene molybdenum tricarbonyl (4)[32].

The importance of stability of a resulting complex is also illustrated by the reaction of molybdenum or tungsten hexacarbonyl with cyclo-octa-1,3-diene which yields the metal tetracarbonyl complexes of cycloocta-1, 5-diene[36] whereas other catalysts bring about the converse isomerisation. Under the same conditions chromium hexacarbonyl converts cycloocta-1, 5-diene into o-xylene.

$$Mo(CO)_6 \rightarrow$$

$$Mo(CO)_3 \quad (4)$$

Allyl ethers are isomerised by iron carbonyls to vinyl ethers and allylbenzene is isomerised to propenylbenzene[30]. Allyl alcohol is isomerised to propionaldehyde by iron pentacarbonyl[19] or cobalt hydrotetracarbonyl[25]. The isomerisation of epoxides to ketones has been effected using a solution of dicobalt octacarbonyl in methanol, which probably contains the species $[Co[CH_3OH)_6][Co(CO)_4]_2$[17]. Propylene oxide, 1, 2-butylene oxide and cyclohexene oxide are converted into acetone, butan-2-one and cyclohexanone respectively. Substitution of pyridine for methanol gives an inactive catalyst. The most remarkable feature of this rearrangement is the formation of ketones rather than the aldehydes which normally result from such reactions.

Very little is known about the mechanism of metal carbonyl catalysed hydrogen migration. Migrations catalysed by iron carbonyls have been discussed in terms of the hydride addition-elimination mechanism[37]. The observation that ketones, ethers, alcohols and especially pyridine promote the catalytic activity and that the resulting isomerisation mixtures have the characteristic red colour of iron carbonyl anion support such a mechanism. The isomerisation of but-1-ene by $[DFe(CO)_4]^-$ in deuterium oxide shows a similar hydrogen-deuterium exchange pattern to that observed for rhodium- and platinum-catalysed reactions[13a]. The isomerisation to deuteration ratio indicates that olefin exchange is fast relative to loss of proton.

The same type of mechanism has been considered for double-bond migration catalysed by cobalt hydrotetracarbonyl[31]. Isomerisation of allylbenzene to propenylbenzene by both cobalt hydrotetracarbonyl and deuterotetracarbonyl proceed at the same rate[42]. The lack of an isotope effect accords with a rate determining step for the formation of $[(olefin)Co(CO)_3D]$. However, the minor incorporation of deuterium which was observed can be explained if only a minor portion of the cobalt hydrotetracarbonyl is involved in the isomerisation sequence as a result of faster exchange of olefin with olefin complex than with cobalt hydrotetracarbonyl. Isomerisation of allyl alcohol to propionaldehyde with cobalt deuterotetracarbonyl results in deuterium labelling solely at the β-carbon atom[25].

$$CH_2 = CH.CH_2OH \xrightarrow{DCo(CO)_4} D.CH_2.CH_2.CHO$$

Transition Metal Intermediates

Reaction of cobalt deuterocarbonyl with penta-1,4-diene produces the complex (5) as a result of a 1,2-hydrogen shift[29].

(5)

Isomerisation of olefins has recently been observed using catalyst systems produced by reaction of triethylaluminium with an iron, cobalt, nickel or palladium salt or acetylacetonate[11a, 20, 35, 45, 48] The addition of 2 or more moles triphenylphosphine per mole nickel suppresses catalytic activity. As expected terminal olefins are converted to internal ones and cycloocta-1,5-diene is converted to cycloocta-1,3-diene. The most interesting case is the isomerisation of cis, trans-cyclodeca-1,5-diene (6) to cis, cis-cyclodeca-1,6-diene (7) accompanied by some 1,2-divinylcyclohexane resulting from thermal rearrangement[45]. Examination of models shows that the diene (7) unlike (6) can adopt a conformation (8) in which the two double bonds are ideally situated for coordination to a single metal atom.

(6)   (7)   (8)

Another synthetically useful isomerisation which involves Ziegler type species is the conversion of isopropylmagnesum bromide to n-propylmagnesium bromide catalysed by the addition of a small amount of titanium tetrachloride[12a]. The reaction almost certainly entails the elimination of a titanium hydride species from an isopropyltitanium compound followed by its addition to the propylene in the opposite sense. As anticipated, addition of another olefin, pent-1-ene, leads to the formation of n-amylmagnesium bromide in reasonable yield. However, no exchange could be detected with either pent-2-ene or cyclohexene. The sequence observed has been depicted as follows:-

The importance of the presence of a β-hydrogen in the Grignard reagent is shown by the lack of exchange between methylmagnesium bromide and pent-1-ene. The same system using isopropylmagnesium bromide catalyses the slow isomerisation of oct-1-ene to *cis* and *trans*-oct-2-enes.

TABLE 3.1 Isomerisation of Olefins

Olefin	Catalyst	Product	References
But-1-ene	$(Et_3PO)_4Ni$	But-2-enes	13a
	$[py_2NiCl_2] + Et_3Al$	But-2-enes	11a
	$CrCl_3 + Et_3Al$	cis and trans-but-2-enes	20
	$PdCl_2$	cis and trans-but-2-enes	13a, 21, 38
	H_2PtCl_6	But-2-enes	13a, 21, 22
	$RhCl_3$	But-2-enes	13, 13a
Pent-1-ene	$PdCl_2$	cis and trans-pent-2-ene	8
	$H_2PtCl_6 + SnCl_2$	cis and trans-pent-2-ene	9
cis-Penta-1,3-diene	$Fe(CO)_5$	trans-penta-1,3-diene $Fe(CO)_3$	18
Penta-1,4-diene	$Fe(CO)_5$	trans-penta-1,3-diene $Fe(CO)_3$	18, 33
Hex-1-ene	$Fe_3(CO)_{12}$	Hexenes (Δ^1 1·6%; cis Δ^2 16%; trans Δ^2 58%; Δ^3 25%)	37
	Fe ethylhexanoate + Et_3Al	Hexenes (cis Δ^2 21·5%; cis Δ^3 and trans Δ^2 72%)	35
	$RhCl_3$	Hexenes (Δ^1 and trans Δ^3 21%; cis Δ^2 21%; trans Δ^2 + cis Δ^3 63%)	27
	$PdCl_2(C_6H_5CN)_2$	Hexenes	27
	$[(C_2H_4)PtCl_2]_2$	Hexenes	27
cis-Hex-2-ene	$Fe_3(CO)_{12}$	Hexenes (cis Δ^2 74%; trans Δ^2 + Δ^3 25%)	37
2-Methylpent-1-ene	$PdCl_2$	2-Methylpentenes (Δ^1 13%; Δ^2 74%; cis Δ^3 2%; trans Δ^3 10%; Δ^4 1%)	5

Substrate	Catalyst	Product	Ref.
4-Methylpent-1-ene	$PdCl_2$ or $PdCl_2(C_6H_5CN)_2$	2-Methylpentenes (Δ^1 13%; Δ^2 73%; cis Δ^3 2%; $trans$ Δ^3 11%; Δ^4 1%)	5, 10, 44, 46
	$[py_2NiCl_2] + Et_3Al$	2-Methylpentenes	11a
2-Methylpent-2-ene	$Fe_3(CO)_{12}$	Unchanged	37
2-Methylpent-3-ene	$Fe_3(CO)_{12}$	2-Methylpentenes (Δ^3 12%; Δ^1 12%; Δ^2 77%)	37
	$[py_2NiCl_2] + Et_3Al$	2-Methylpentenes	11a
$trans$-2-Methylpent-3-ene	$Fe_3(CO)_{12}$	2-Methylpentenes (Δ^3 95%; Δ^1 2%; Δ^2 2%)	37
Hexa-1,5-diene	$Fe(CO)_5$	$trans$-Hexa-1,3-diene $Fe(CO)_3$	2, 18
$trans,trans$-Hexa-2,4-diene	$Fe(CO)_5$	$trans,trans$-Hexa-2,4-diene $Fe(CO)_3$	18
$cis,trans$-Hexa-2,4-diene	$Fe(CO)_5$	$cis,trans$-Hexa-2,4-diene $Fe(CO)_3$	18
Cyclohexa-1,4-diene	$Fe(CO)_5$	Cyclohexa-1,3-diene $Fe(CO)_3$	18
Hept-2-ene (cis and $trans$)	$RhCl_3$	Heptenes (Δ^1 + $trans$ Δ^3 30%; $trans$ Δ^2 + cis Δ^3 54%; cis Δ^2 16%)	27
2-Methylhexa-1,5-diene	$Fe(CO)_5$	$trans$-2-Methylhexa-1,3-diene $Fe(CO)_3$	18
3 and 4-Methylcyclohexenes	$Fe_3(CO)_{12}$	Unchanged	37
Oct-1-ene	$Fe_3(CO)_{12}$	Octenes ($trans$ Δ^4 19%; cis Δ^4 2%; $trans$ Δ^3 33%; cis Δ^3 6%; $trans$ Δ^2 32%; cis Δ^2 7%)	7, 11
	$Mo(CO)_6$	Octenes	37
	$RhCl_3$	Octenes (similar composition to above)	5
	$PdCl_2$	Octenes (similar composition to above)	5, 14
	$PtCl_2$	Octenes (similar composition to above)	5
	H_2PtCl_6	Octenes	21, 22

(Continued overleaf)

TABLE 3.1 (continued)

Olefin	Catalyst	Product	References
Oct-1-ene	$[IrCl_3P_3]$ where $P = Et_3P$, Bu_3P, Et_2PhP	Oct-2-ene and a little oct-3-ene	12
	$[IrH_3(Ph_3P)_3]$	Octenes	12
	$[IrHCl_2(Et_2PhP)_3]$	Octenes	12
trans-Oct-4-ene	$PdCl_2$	Octenes (composition as for oct-1-ene)	5
2, 5-Dimethylhexa-1, 5-diene	$Fe(CO)_5$	*trans*-2, 5-Dimethylhexa-1, 3-diene $Fe(CO)_3$	18
2, 5-Dimethylhexa-2, 4-diene	$Fe(CO)_5$	2, 5-Dimethylhexa-1, 3-diene $Fe(CO)_3$	18
Vinylcyclohexane	$Fe_3(CO)_{12}$	Ethylidene cyclohexane and ethylcyclohex-1-ene	37
Cycloocta-1, 4-diene	$Fe(CO)_5$	Cycloocta-1, 3-diene 88%	48
Cycloocta-1, 5-diene	$Mn_2(CO)_{10}$	Cycloocta-1, 3-diene 19%	48
	$Fe(CO)_5$	Cycloocta-1, 3-diene 98%	2, 48
	$Co_2(CO)_8$	Cycloocta-1, 3-diene 32%	48
	$Cr(acac)_3 + Et_3Al$	Cycloocta-1, 3-diene 9%	48
	$Fe(acac)_3 + Et_3Al$	Cycloocta-1, 3-diene 66%	48
	Fe ethylhexanoate + Et_3Al	Cyclooctadienes (1, 3 87%; 1, 4 6·5%; 1, 5 6·5%)	35
	$Ni(acac)_2 + Ph_3P + Et_3Al$	Cycloocta-1, 3-diene 85%	48
	$Pd(OAc)_2 + Et_3Al$	Cycloocta-1, 3-diene 36%	48
	$RhCl_3$	Cycloocta-1, 3-diene	40, 41

82

	Catalyst	Product	Reference
	$[RhCl_3(Et_2PhP)_3]$	Cycloocta-1,3-diene	4
	H_2IrCl_6	Cycloocta-1,3-diene	40
	$[IrHCl_2(Et_2PhP)_3]$	Cycloocta-1,3-diene	40
	$[IrCl_3P_3]$ where $P = Et_3P$, Et_2PhP, Me_2PhP	Cycloocta-1,3-diene	40
	$PdCl_2$	Cycloocta-1,3-diene 80%	23, 34
	$PdCl_2(C_8H_{12})$	Cycloocta-1,3-diene	34
	trans $PdI_2(Me_2PhP)_2$	Cycloocta-1,3-diene	40
	trans $PtHCl(Et_3P)_2$	Cycloocta-1,3-diene	40
1,4-Dihydromesitylene	$Fe(CO)_5$	1,2-Dihydromesitylene $Fe(CO)_3$	33
Allylbenzene	$H Co(CO)_4$	Propenylbenzene	42
	$Fe(CO)_5$	Propenylbenzene (trans 95%, cis 5%)	30
	$PdCl_2$	Propenylbenzene	16
Undec-1-ene	$Fe(CO)_5$ + light	Undec-2, 3, 4 and 5-enes	4, 6, 7
	$Co_2(CO)_8$ + light	Undec-2, 3, 4 and 5-enes	6
	$Os_2(CO)_9$ + light	Undec-2, 3, 4 and 5-enes	6
Dodec-1-ene	$Fe(CO)_5$	2, 3, 4, 5 and 6-Dodecenes	3
	$Co_2(CO)_8$	2, 3, 4, 5 and 6-Dodecenes	3
trans, cis-Cyclododeca-1, 5-diene	$Fe(acac)_3$ + Et_2AlOEt	cis, cis-Cyclododeca-1, 6-diene 48%	45
	$Co(acac)_3$ + Et_2AlOEt	cis, cis-Cyclododeca-1, 6-diene 66%	45
	$Ni(acac)_2$ + Et_2AlOEt	cis, cis-Cyclododeca-1, 6-diene 61%, divinylcyclohexane 39%	45

(Continued overleaf)

TABLE 3.1 (continued)

Olefin	Catalyst	Product	References
trans,cis-Cyclododeca-1,5-diene	$Pd(OAc)_2$ + Et_2AlOEt	cis,cis-Cyclododeca-1,6-diene 90%	45
cis-Stilbene	$Fe_3(CO)_{12}$	trans-Stilbene	37
trans-Stilbene	$Fe_3(CO)_{12}$	Unchanged	37
Allylpentafluorobenzene	$Fe(CO)_5$	Propenylpentafluorobenzene (trans 98%, cis 2%)	30
Allyl ethyl ether	$Fe(CO)_5$	Propenyl ethyl ether (trans 56%, cis 44%)	30
But-3-enyl methyl ether	$Fe(CO)_5$	But-1-enyl methyl ether (trans 41%, cis 34%)	30
Diallyl ether	$Fe(CO)_5$	Dipropenyl ether (trans,trans 31%; cis,trans 43; cis,cis 26%)	30
Allyl phenyl ether	$Fe(CO)_5$	Propenyl phenyl ether (trans 48%, cis 52%)	30

REFERENCES

1. Alderson, T., Jenner, E. L. and Lindsey, R. V., J. Amer. Chem. Soc., 1965, **87**, 5638.

2. Arnett, J. E. and Pettit, R., J. Amer. Chem. Soc., 1961, **83**, 2954.

3. Asinger, F. and Berg, O., Chem. Ber., 1955, **88**, 445.

4. Asinger, F., Fell, B. and Collin, G., Chem. Ber., 1963, **96**, 716.

5. Asinger, F., Fell. B. and Krings, P., Tetrahedron Letters, 1966 633.

6. Asinger, F., Fell, B. and Schrage, K., Chem. Ber., 1965, **98**, 372.

7. Asinger, F., Fell. B. and Schrage, K., Chem. Ber., 1965, **98**, 381.

8. Bond. G. C. and Hellier, M., J. Catalysis, 1965, **4**, 1.

9. Bond, G. C. and Hellier, M., Chem. and Ind., 1965, 35.

10. British Petroleum Co. Ltd. Belg. P. 612, 300; Chem. Abs., 1962 **57**, 13605.

11. Carr, M. D., Kane, V. V. and Whiting, M. C., Proc. Chem. Soc., 1964, 408.

11a. Chauvin, Y. and Lefebvre, G., Compt. rend., 1964, **259**, 2105.

12. Coffey, R. S., Tetrahedron Letters, 1965, 3809.

12a. Cooper. G. D. and Finkbeiner, H. L., J. Org. Chem., 1962, **27**, 1493.

13. Cramer, R., J. Amer. Chem. Soc., 1966, **88**, 2272.

13a. Cramer. R. and Lindsey, R. V., J. Amer. Chem. Soc., 1966, **88**, 3534.

14. Davies, N. R., Austral, J. Chem., 1964, **17**, 212.

15. Davies, N. R., Nature, 1964, **201**, 490.

16. Davies, N. R., Nature, 1965, **205**, 281.

17. Eisenmann, J. L., J. Org. Chem., 1962, **27**, 2706.

18. Emerson, G. F., Mahler, J. E., Kochhar, R. and Pettit, R., J. Org. Chem., 1964, **29**, 3620.

19. Emerson, G. F. and Pettit, R., J. Amer. Chem. Soc. 1962, **84**, 4591.

20 Erasova, E. L., Krentsel, B. A., Pokatilo, N. A. and Topchiev, A. V., Vysokomolecul. Soedin., 1962, **4**, 1796; Chem. Abs. 1963, **59**, 765.

21. Feller, M., U.S.P.2, 960, 551; Chem. Abs., 1961, **55**, 5344.

22. Feller, M., Brennan, H.M. and Seelig, H.S., U.S.P.2, 960, 550; Chem. Abs., 1961, **55**, 5344.

23. Frye, H., Kuljian, E. and Viebrock, J., Inorg. Chem., 1965, **4**, 1499.

24. Frye, H., Kuljian, E. and Viebrock, J., Inorg. Nucl. Chem. Letters, 1966, **2**, 119.

25. Goetz, R. W. and Orchin, M., J. Amer. Chem. Soc., 1963, **85**, 1549.

26. Haddad, Y. M. Y., Henbest, H. B., Husbands, J. and Mitchell, T. R. B., Proc. Chem. Soc., 1964, 361.

27. Harrod, J. F. and Chalk, A. J., J. Amer. Chem. Soc., 1964, **86**, 1776.

28. Harrod, J. F. and Chalk, A. J., Nature., 1965, 205, 280.

29. Husebye, S., Jonassen, H. B. and Moore, D. W., Acta Chem. Scand., 1964, **18**, 1581.

30. Jolly, P. W., Stone, F. G. A. and Mackenzie, K., J. Chem. Soc., 1965, 6416.

31. Karapinka, G. L. and Orchin, M., J. Org. Chem., 1961, **26**, 4187.

32. King, R. B. and Stone, F. G. A., Chem. and Ind., 1960, 232.

33. King, R. B., Manuel, T. A. and Stone, F. G. A., J. Inorg. Nucl. Chem., 1961, **16**, 233.

34. Lafont, P. and Vivant, G., Fr. P. 1, 337, 889; Chem. Abs., 1964, **60**, 2802.

35. Lapporte, S. J., Fr. P. 1, 403, 409; Chem. Abs., 1965, **63**, 14725.

36. Leigh, G. J. and Fischer, E. O., J. Organometallic Chem., 1965, **4**, 461.

37. Manuel, T. A., J. Org. Chem., 1962, **27**, 3941.

38. Moiseev, I. I. and Pestrikov, S. V., Izv. Akad. Nauk S.S.S.R., Ser. Khim., 1965, 1717.

39. Nicholson, J. K. and Shaw, B. L., Proc. Chem. Soc., 1963, 282.

40. Nicholson, J. K. and Shaw, B. L., Tetrahedron Letters, 1965, 3533.

41. Rinehart, R. E. and Lasky, J. S., J. Amer. Chem. Soc., 1964, **86**, 2516.

42. Roos, L. and Orchin, M., J. Amer. Chem. Soc., 1965, **87**, 5502.

43. Sarel, S., Ben-Shoshan, R. and Kirson, B., J. Amer. Chem. Soc., 1965, **87**, 2517.

44. Sparke, B., Turner, L. and Wenham, A. J. M., J. Catalysis., 1965, **4**, 332.

45. Studiengesellschaft m.b.H., Fr.P.1, 386, 991; Chem. Abs., 1965, **62**, 16085.

46. Turner, L., B. P. 932, 748; Chem. Abs., 1964, **60**, 405.

47. Vaska, L. and Di Luzio, J. W., J. Amer. Chem. Soc., 1962, **84**, 4989.

48. Wittenberg, D. and Seibt, H., G. P.1, 136, 329; Chem. Abs., 1963, **58**, 4442.

49. Young, J. F., Gillard, R. D. and Wilkinson, G., J. Chem. Soc., 1964, 5176.

CHAPTER 4

THE OXIDATION OF OLEFINS BY PLATINUM METAL COMPOUNDS

It has been known for a long time than an aqueous solution of palladous chloride can oxidise ethylene to acetaldehyde with concomitant deposition of palladium [80]. The hydrolysis of Zeise's salt. $K[(C_2H_4) PtCl_3]$, produces ethylene, acetaldehyde and ethanol accompanied by deposition of platinum. Initially, the formation of acetaldehyde was attributed to oxidation of ethanol by platinum (II) and only recently has it been suggested that it results from the initial breakdown of the complex[60]. The accidental rediscovery of these reactions in the laboratories of the Consortium für Elektrochemische Industrie and development by the parent company, Wacker-Chemie, has resulted in the Wacker Process for converting ethylene into acetaldehyde[89, 92].

The overall process is summarised by the equation:-

$$C_2H_4 + Pd\ Cl_2 + H_2O \rightarrow CH_3CHO + Pd + 2HCl.$$

Other olefins are oxidised to ketones. A number of kinetic studies [26, 42, 43, 59a, 71, 75, 106, 107] have established that the equation for the rate of oxidation of olefins by palladous chloride in aqueous solution is

$$-\frac{d[olefin]}{dt} = \frac{Kk[olefin][Pd\ Cl_4{}^{2-}]}{[Cl^-]^2[H^+]}$$

although the rate of oxidation of cyclohexene is apparently independent of acid concentration[105]. It has also been confirmed that the oxidation does not proceed through hydration and subsequent oxidation of the alcohol. Although palladous chloride oxidises alcohols to aldehydes the reaction is much slower than olefin oxidation [37, 76].

The mechanism of olefin oxidation has been widely discussed[42, 59a, 65, 70, 71, 72, 73, 75, 91, 106, 107, 108]. It is generally agreed that the first step is coordination of the olefin.

$$Pd\ Cl_4{}^{2-} + olefin \overset{K}{\rightleftharpoons} [(olefin)Pd\ Cl_3]^- + Cl^- \qquad (i)$$

The values of the equilibrium constant K, which appears in the rate equation, for a series of olefins parallel the equilibrium constants for their complexation with silver ion[43].

The first mechanism to be proposed for the conversion of the above complex to products involved a nucleophilic attack by hydroxide ion on one of the olefinic carbon atoms accompanied by a simultaneous 1, 2 hydrogen shift and expulsion of the palladium, equation (ii)[37, 89, 92].

$$\left[\begin{array}{c} H \\ C = C \\ H \quad \downarrow \quad H \\ PdCl_3 \end{array} \right]^{\ominus} + OH^{\ominus} \longrightarrow CH_3 - C \overset{\overset{\oplus}{O}-H}{\underset{H}{\diagdown}} + Pd + 3\,Cl^{\ominus} \qquad (ii)$$

A principle objection to this mechanism is the difficulty in accounting for the inhibition of complex decomposition by acid. This can hardly be ascribed to a decrease in the concentration of the hydroxide ion as, under typical contions employing decinormal acid, the concentration of hydroxide ion is only 10^{-13}M so that nucleophilic attack by water would be far more important.

The most widely accepted mechanism is summarised in equations (iii) to (vi)[42, 43, 59a].

$$[(\text{olefin})PdCl_3]^- + H_2O \rightleftharpoons [(\text{olefin})PdCl_2(H_2O)] + Cl^- \qquad (iii)$$

$$[(\text{olefin})PdCl_2(H_2O)] + H_2O \rightleftharpoons [(\text{olefin})PdCl_2(OH)]^- + H_3O^+ \quad (iv)$$

$$[(\text{olefin})PdCl_2(OH)]^- + H_2O \xrightarrow{\text{slow}} \left[\begin{array}{c} R_3 \quad R_2 \\ | \quad\quad | \\ HO - C - C - PdCl_2(H_2O) \\ | \quad\quad | \\ H \quad R_1 \end{array} \right] \quad (v)$$

$$\downarrow \text{fast}$$

$$R_1R_2CH.CO.R_3 + HCl + Cl^- + Pd \quad (vi)$$

The inhibition by acid obviously operates on step (iv), where the loss of a proton is in keeping with the increase in acidity of a water molecule on complexing, by a factor of 10^7 to 10^{10}. Measurement of the second order rate constant (k_2) for the oxidation of ethylene in water and deuterium oxide[73] leads to a value for the isotope effect $(k_2^{H_2O}/k_2^{D_2O})$ of 4. 05. This is of the order of the decrease in ionisation constants of weak acids in deuterium oxide. The equilibria in equations (i) and (iii) accommodate the inverse second order dependence on chloride ion concentration.

Transition Metal Intermediates

Although the hydroxo complex could be envisaged as decomposing directly to product experimental evidence indicates that the breakdown occurs by way of a σ-complex, equations (v) and (vi). Decomposition of the binuclear ethylene-palladous chloride complex (1) to acetaldehyde with deuterium oxide does not result in deuterium incorporation [37, 92], hence it must be concluded that all of the hydrogen atoms of the acetaldehyde are derived from ethylene. Apart from excluding a vinyl alcohol intermediate this also means that decomposition of the hydroxo complex is accompanied by a hydride shift. If this were to occur as part of a one step decomposition process a primary isotope effect should be observed for the oxidation of tetradeuteroethylene, whereas only a secondary one of 1.07 is found[42].

(1)

The formation of the σ-complex is consistent with known reactions of palladous chloride with olefins forming β-alkoxyalkylpalladous chlorides [13]. Further, the reaction of palladous chloride with appropriate mercuric chloride derivatives, which would be expected to generate transiently the appropriate σ-complexes, produces the observed end products [69].

$$HOCH_2CH_2HgCl + PdCl_2 \rightarrow CH_3CHO + Pd + HgCl_2 + HCl$$

$$C_2H_5.O.CH_2CH_2HgCl + PdCl_2 \rightarrow C_2H_5.O.CH = CH_2 + Pd + HgCl_2 + HCl$$

$$CH_3.CO.O\ CH_2CH_2HgCl + PdCl_2 \rightarrow CH_3.CO.O\ CH = CH_2 + Pd + HgCl_2 + HCl$$

The actual decomposition is pictured as shown in (2), with the departing palladium assisting the hydride shift[42].

(2)

In practice the oxidation of olefins is generally conducted by using a small amount of a palladous salt together with an oxidising agent capable of reoxidising palladium (0) to palladium (II). A wide range of oxidants have been employed including benzoquinone, ferric salts, potassium dichromate, potassium persulphate, potassium bromate and lead dioxide. Electrolytic oxidation has also been used[34, 62].

However, the most common oxidant employed is a cupric salt either by using a quantitative amount, or a lesser amount together with a supply of oxygen capable of reoxidising the cuprous ion. Direct aereal oxidation of palladium to palladium (II) is too slow. An additional advantage of using a copper salt is that copper (I) forms olefin complexes thereby increasing the amount of olefin available in solution particularly as these complexes are less stable than the palladium ones.

In reactions using catalytic amounts of both palladium and copper salts the oxidation of palladium by copper (I) can limit the reaction rate. The addition of a ligand capable of forming a bridge between palladium and copper improves the reaction rate [65, 79]. Ligands can be arranged in order of increasing accelerating effect:- salicylate < malate, acetate, nitrate, citrate, phthalate < tartrate < ethylaceto-acetate < acetylacetone[65]. Although the oxidation of copper (I) is not often rate limiting the process can be speeded up by addition of potassium naphtho-1, 2-quinone-4-sulphonate[24].

Anions can also influence the reaction in other ways. The rate equation shows the rate of reaction to be inversely proportional to the square of the chloride ion concentration. The chloride ion suppresses complex formation between the olefin and palladium (II). Anions such as sulphate, nitrate, phosphate, perchlorate and fluoride which have very limited coordinating ability have little effect on the reaction rate. However chloride, bromide and iodide ions, whose coordinating power increases in this order, show corresponding inhibitory effects[92]. The optimum copper to chloride ratio lies between 1:1·4 and 1:1·8. [84] The principal byproducts from the oxidation of ethylene are chlorinated compounds including chloroform, methylene chloride, ethyl chloride, 1, 2-dichloroethane, chloroacetaldehydes and 2-chloroethanol. Formation of byproducts can be reduced by avoiding the use of halide ions and strong mineral acids[10].

Olefins of commercial origin sometimes contain small amounts of acetylenic impurities which inactivate the palladium catalyst. This difficulty can be circumvented by adding a source of mercuric ions to the oxidising medium[20]. The presence of cadmium, zinc or aluminium ions is reported to increase yields[32, 33, 56] but the reason for this is not apparent. Palladium salts appear to be the most effective catalysts for olefin oxidation, but platinum, rhodium, iridium, osmium and ruthenium salts have also been used[7, 21, 28, 83, 89, 90, 91].

Oxidation of olefins other than ethylene produces mainly ketones (Table 4. 1). Terminal olefins produce mostly methyl ketones with minor amounts of aldehydes. The proportion of aldehyde depends on the reaction temperature, the pH of the solution and the palladium salt employed[37, 83]. As noted earlier the rate of oxidation is closely related to the coordinating ability of the olefin[43, 89], and the struc-

TABLE 4.1 Oxidation of Olefins to Carbonyl Compounds

Olefin	Products	References
Ethylene	Acetaldehyde (90-98%)	7, 8, 10, 11, 18, 24, 28, 29, 30, 31, 32, 33, 38, 39, 44, 49, 56, 58, 62, 66, 82, 85, 90, 93, 98, 101
Propylene	Acetone and a little propionaldehyde	10, 21, 28, 30, 37, 83, 90, 95, 98
But-1- and -2-ene	Methyl ethyl ketone (85-88%), butyraldehyde (2-4%)	18, 28, 30, 34, 37, 56, 79, 95
Isobutene	Methacrolein (36%), iso-butyraldehyde (12%)	2, 46, 47, 48
	t-Butanol, iso-butyraldehyde	28
	t-Butanol, acetone	91
Buta-1,3-diene	Crotonaldehyde (>90%), diacetyl	30, 90, 91
Pent-1-ene	Pentan-2-one (>90%)	37, 90
Pent-2-ene	Pentan-2-one (>90%)	90
Penta-1,4-diene	Pent-2-en-1-al (>90%)	90
Cyclopentene	Cyclopentanone (>90%)	90
2-Methylbut-1-ene	α-Methylcrotonaldehyde (18%), 2-methylbutyraldehyde	46

Olefin	Product(s)	References
2-Methylbut-2-ene	α-Methylcrotonaldehyde (19%), 2-methylbut-1-en-3-one (2%)	2, 3, 47
	3-Methylbut-2-enal	47
	α-Methylcrotonaldehyde (19%), 2-methylbutan-3-one (2%)	46
	α-Methylcrotonaldehyde, 2-methylbut-1-ene-3-one	3
	3-Methylbut-2-enal	3
Hex-1-ene	Hexan-2-one (>90%)	37, 90
Cyclohexene	Cyclohexanone	1, 10, 48, 90
2-Methylpent-1-ene	Mesityl oxide (20%)	46
2-Methylpent-2-ene	Mesityl oxide (19%), 2-methylpentan-3-one (1%)	46
3-Methylpent-2-ene	3-Methylpent-2-en-4-one (62%)	3, 46
4-Methylpent-1-ene	2-Methylpentan-4-one	12, 25, 40
2,2-Dimethylbut-3-ene	Pinacolone (66%)	48
Hept-1-ene	Heptan-2-one (>90%)	37, 74, 90
3-Ethylpent-2-ene	3-Ethylpent-2-en-4-one	2, 47
1-Methylcyclohexene	Methylcyclohexane and toluene	48
Oct-1-ene	Octan-2-one (>90%)	90
Non-1-ene	Nonan-2-one (>90%)	19, 90
1-Methylcyclooctene	Formylcyclooct-1-ene	2, 47

(Continued overleaf)

93

TABLE 4.1 (continued)

Olefin	Products	References
Dec-1-ene	Decan-2-one (>90%)	90
Dodec-1-ene	Dodecan-2-one (>80%)	17
Styrene	Acetophenone (>90%)	56, 90
	Phenylacetaldehyde, acetophenone	37
m-Nitrostyrene	m-Nitroacetophenone (35%)	97
α-Methylstyrene	Acetophenone, 2,4-diphenyl-4-methylpent-2-ene	48
β-Methylstyrene	Methyl benzyl ketone	48
Allylbenzene	Methyl benzyl ketone (>90%)	90
Indene	β-Indanone (>90%)	90
α,β-Dimethylstyrene	2-Phenylbutan-3-one, 2,4-diphenyl-3,4-dimethylhex-2-ene	48
Stilbene	Desoxybenzoin (92%)	48
1,1-Diphenylethylene	Benzophenone, 1,1,4,4-tetraphenylbutadiene	48
Vinyl chloride	Acetaldehyde	91
Allyl chloride	Methylglyoxal (65%)	48
1,3-Dichloropropene	Methylglyoxal (32%)	91
Methallyl chloride	α-Methylacrolein (49%), iso-butyraldehyde (16%)	48

Crotonaldehyde	β-Oxobutyraldehyde	91
Undec-10-enoic acid	10-Ketoundecanoic acid (77%)	17
Acrylonitrile	Acetylcyanide (88%)	97
Crotonamide	Acetone (80%)	97
α-Methylacrylamide	Propionaldehyde (82%)	97
β-Ethylacrylamide	Methyl ethyl ketone (70%)	97
β-n-Propylacrylamide	Pentan-2-one (68%)	97
β-n-Butylacrylamide	Hexan-2-one (53%)	97
Cinnamamide	Acetophenone (48%)	97
m-Nitrocinnamamide	m-Nitroacetophenone (38%)	97
Allylamine	Propionaldehyde (15%), methylglyoxal (1%)	97
	Acetaldehyde (30%), methylglyoxal (7%)	97
α-Methylallylamine	Methyl ethyl ketone (45%), diacetyl (13.5%)	97
Crotyldiethylamine	Methyl ethyl ketone (36%), unidentified carbonyl compounds (50%)	97
Allylurea	Propionaldehyde (24%)	97
Nitroethylene	α-Nitroacetaldehyde (5.5%)	97
1-Nitropropyne	α-Nitroacetone (37%)	97

TABLE 4.2 Oxidation of π-Allylpalladium Chloride Complexes[46]

Parent Olefin	Method*	Products	Yield, %
Isobutene	B	Methacrolein	40
2-Methylbut-1 or 2-ene	C	α-Methylcrotonaldehyde and methyl isopropenyl ketone	57
	D	β-Methylcrotonaldehyde	37
	B	α-Methylcrotonaldehyde and methyl isopropenyl ketone (1:1)	84
2-Methylpent-1 or 2-ene	D	Mesityl oxide	28
	A	2-Methylpenten-3-one	26
	B	2-Methylpent-2-enal and 2-Methylpenten-3-one	18
3-Methylpent-2-ene	A	3-Methylpent-2-en-4-one	29
	B	3-Methylpent-2-en-4-one	38
2,3-Dimethylbut-1 or 2-ene	B	2,3-Dimethylbut-2-enal	13
3-Ethylpent-2-ene	C	3-Ethylpent-2-en-4-one	trace
	D	3-Ethylpent-2-en-4-one	trace
	A	3-Ethylpent-2-en-4-one	74
2,4-Dimethylpent-2-ene	B	2,4-Dimethylpent-2-enal	28
2,4,4-Trimethylpent-2-ene	A	2,4,4-Trimethylpent-2-enal	19

* A - sodium dichromate and sulphuric acid;   B - manganese dioxide in sulphuric acid;
C - unbuffered palladous chloride;   D - buffered palladous chloride.

96

ture of the olefin seems to have a relatively small effect on the subsequent stages. In general olefins with internal double-bonds are oxidised less rapidly than terminal ones, and *cis*-isomers are more rapidly oxidised than *trans* ones. The completely aqueous reaction systems used for oxidation of ethylene are less satisfactory for oxidation of higher molecular weight olefins. The best medium in such cases is aqueous dimethylformamide. Other cosolvents such as dimethylsulphoxide, acetone, acetic acid, tetrahydrofuran, dioxan and acetonitrile are markedly inferior[17].

A variety of side-reactions are encountered. Most, such as the hydrolysis and decarboxylation which accompany the oxidation of $\alpha\beta$-unsaturated carboxylic acid derivatives are unexceptional, e.g.

$$CH_3.CH = CH.CO.NH_2 \longrightarrow [CH_3.CO.CH_2CONH_2] \longrightarrow CH_3CO.CH_3$$

Also the double-bond migration which accompanies the oxidation of butadiene to crotonaldehyde is not surprising:

$$CH_2 = CH.CH = CH_2 \longrightarrow CH_2 = CH.CH_2.CHO$$
$$\longrightarrow CH_3CH = CH.CHO$$

Unsymmetrically disubstituted olefins are especially prone to produce products resulting from side-reactions[48]. In order to minimise acid catalysed addition to the double-bond the oxidation is frequently carried out in the presence of acetate buffer, which also favours the formation of π-allylpalladium complexes. The oxidation of α-methylstyrene with palladous chloride in aqueous acetic acid exemplifies the types of product encountered: [48]

In the presence of sodium acetate the oxidation also produces 2, 5-diphenylhexa-2, 4-diene. The mode of formation of these products is unknown. The cleavage of the double bond also occurs to a minor extent with other olefins and is favoured by low concentrations of hydrogen and chloride ions.

The oxidation of methyl or methylene groups adjacent to double bonds is most frequently encountered with olefins having di- or trisubstituted double bonds. These products may arise by oxidation of an intermediate π-allylpalladium complex, since the same products are obtained from decomposition of the complex as are produced by

97

direct odixation of the olefin. The factors determining the site of
oxidation are not clear. In the case of the oxidation of 2-methylbut-1
or 2-ene the pH of the solution controls the products formed[46]. In
50% aqueous acetic acid the product is α-methylcrotonaldehyde,
while if the pH is adjusted to 3 β-methylcrotonaldehyde is formed.

$$
\begin{array}{c}
CH_3 \\
| \\
CH_2=C\cdot CH_2\cdot CH_3 \\
CH_3 \\
| \\
CH_3\cdot C=CH\cdot CH_3
\end{array}
\quad
\begin{array}{l}
\xrightarrow{50\%\ AcOH} \quad CH_3\cdot CH=\overset{\overset{\textstyle CH_3}{|}}{C}\cdot CHO \\
\\
\xrightarrow{pH3} \quad CH_3\cdot \overset{\overset{\textstyle CH_3}{|}}{C}=CH\cdot CHO
\end{array}
$$

The same dichotomy is observed when the derived π-allyl complex is
oxidised with palladous chloride under the same conditions. The
oxidation of π-allylpalladium complexes with sodium dichromate or
manganese dioxide in sulphuric acid gives rather better yields of
α, β-unsaturated carbonyl compounds (Table 4. 2)[46]. As the complexes can be obtained from the reaction of palladous chloride with
either olefins or allyl compounds the method is synthetically useful.

As mentioned earlier alcohols are oxidised by palladium (II) to
carbonyl compounds, but allyl alcohol yields propylene, 4-methyl-
enetetrahydrofurfuryl alcohol and 4-methyl-2, 5-dihydrofurfuryl
alcohol[38a]. A mechanism based on that for the oxidation of olefins
has been suggested for the formation of these products.

The first mechanism proposed for oxidation of olefins by palladous
chloride, which postulated a nucleophilic attack by hydroxyl ion on
the coordinated double bond, prompted the examination of the behaviour of other nucleophiles[68, 100]. The ethylenepalladium chloride
complex was found to react with sodium acetate in acetic acid
yielding vinyl acetate and palladium (Figure 4. 1). Reaction with
alcohols gave a mixture of the appropriate vinyl ether and acetaldehyde acetal. Further, reaction with acetamide and subsequent
hydrogenation produced N-ethylacetamide. Although the original

mechanistic concepts are no longer accepted these vinylation reactions have been further developed, mainly as a route to vinyl esters. (Table 4. 3).

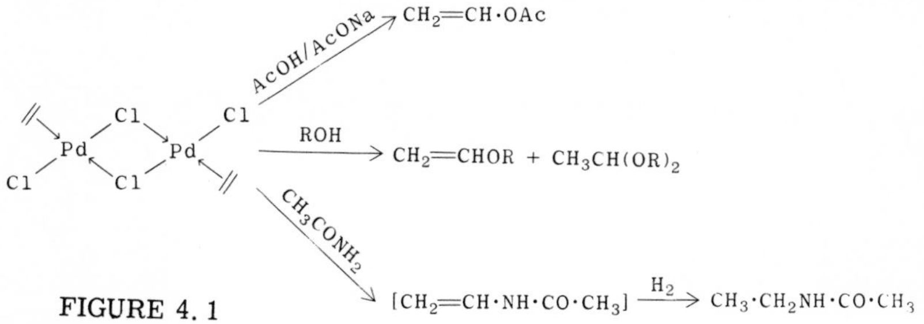

FIGURE 4. 1

These vinylation reactions are generally carried out by using a catalytic amount of a palladium salt and an oxidising agent. The principal oxidants employed are cupric acetate, oxygen and cupric acetate, and benzoquinone. The reaction is first order in ethylene and in palladous acetate[5, 77]. While palladium salts are normally used as catalysts the use of platinum, rhodium or nickel salts has been reported[64]. Even under rigorously anhydrous conditions the formation of vinyl acetate is always accompanied by production of acetaldehyde and ethylidene diacetate. The latter product arises from addition of acetic acid to vinyl acetate and is stable under the reaction conditions. The acetaldehyde is formed by reaction of acetic acid with vinyl acetate, acetic anhydride being the other product. This reaction is specifically catalysed by palladous chloride[16].

An obvious extension of the reaction is to use other carboxylic acids in place of acetic acid and so obtain the appropriate vinyl ester. Alternatively it is possible to effect a transvinylation reaction between the carboxylic acid and vinyl acetate[57, 92, 96].

$$CH_2 = CHO.CO.CH_3 + R.CO_2H \rightleftharpoons CH_2 = CH.O.CO.R + CH_3CO_2H$$

This process is also catalysed by palladium, platinum and rhodium salts, the former being the most effective[92]. The conversion of vinyl chloride into vinyl acetate[81, 86] under these conditions is probably an extension of the reaction. Palladous chloride also accelerates the hydrolysis of vinyl esters.

The products obtained from olefins other than ethylene depend on the olefin, reaction temperature and solvent system. Oxidation of terminal olefins in acetic acid gives mainly enol acetate while other olefins give mainly allylic acetate[61]. Different allylic acetates are obtained from Δ^1 and Δ^2 olefins, cf. Figure 4. 2. It seems that allylic acetates are not formed by acetolysis of a π-allylpalladium complex, as the rates and products of the acetolysis of π-crotylpalladium

TABLE 4.3 Oxidation of Olefins to Vinyl and Allyl Esters

Olefin	Carboxylic Acid	Products	References
Ethylene	Formic acid	Vinyl formate	64
	Acetic acid	Vinyl acetate and a little acetaldehyde	9, 14, 22, 27, 35, 41, 45, 50, 51, 52, 53, 57, 59, 63, 64, 78, 87, 88, 100, 109, 110, 111
	Propionic acid	Vinyl propionate	64
	Butyric acid	Vinyl butyrate	57
	Isobutyric acid	Vinyl *iso*butyrate	35
	Crotonic acid	Vinyl crotonate	57
	Hexanoic acid	Vinyl hexanoate	64
	Benzoic acid	Vinyl benzoate	57
Propylene	Acetic acid	*cis* and *trans*-Propenyl, *iso*-propenyl allyl and *iso*propyl acetates	6, 50, 54, 64, 67, 94, 99
		1,3-Diacetoxypropane	51
	Propionic acid	Isopropenyl propionate	64
But-1-ene	Acetic acid	But-1-en-3-yl, but-1-en-2-yl, *trans* and *cis*-but-2-en-2-yl, *trans* and *cis*-but-2-en-1-yl acetates	4, 51, 61
cis or *trans*-But-2-ene	Acetic acid	But-1-en-3-yl acetate	51, 61

iso-Butylene	Acetic acid	Methallyl acetate	45
Pent-1-ene	Acetic acid	Pent-1-en-2-yl and pent-2-en-1-yl acetates	61
cis-Pent-2-ene	Acetic acid	Pent-1-en-3-yl and pent-3-en-2-yl acetates	61
Hex-1-ene	Acetic acid	Hex-2-en-1-yl acetate (68%)	108
	Propionic acid	Hexenyl propionate (37%)	108
Cyclohexene	Acetic acid	Cyclohexenyl acetate	1
Hept-1-ene	Acetic acid	Hept-2-en-1-yl acetate (66%)	108
Oct-1-ene	Acetic acid	Octen-1-yl and oct-2-enyl acetates (1.9:1; 35%)	14
Oct-2-ene	Acetic acid	*s*-Octenyl acetates	15
Dec-1-ene	Acetic acid	Dec-2-en-1-yl acetate (53%)	108
Allyl acetate	Acetic acid	1,1-Diacetoxyprop-2-ene, propen-1-yl acetate and acraldehyde	55

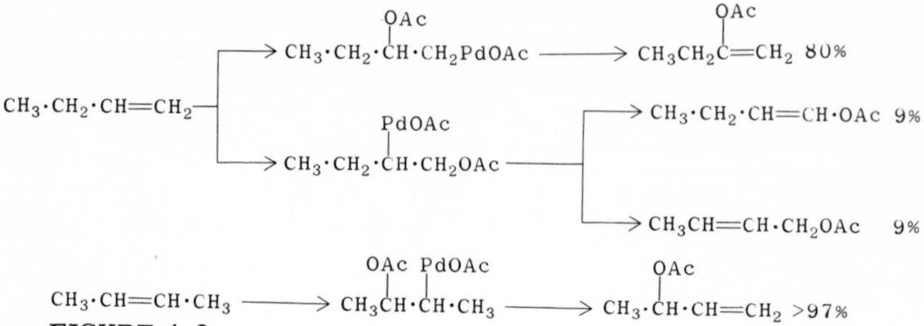

$$CH_3 \cdot CH_2 \cdot CH{=}CH_2 \begin{cases} \longrightarrow CH_3 \cdot CH_2 \cdot \overset{OAc}{\underset{|}{CH}} \cdot CH_2 PdOAc \longrightarrow CH_3 CH_2 \overset{OAc}{\underset{|}{C}}{=}CH_2 \quad 80\% \\ \\ \longrightarrow CH_3 \cdot CH_2 \cdot \overset{PdOAc}{\underset{|}{CH}} \cdot CH_2 OAc \longrightarrow \begin{cases} \longrightarrow CH_3 \cdot CH_2 \cdot CH{=}CH \cdot OAc \quad 9\% \\ \\ \longrightarrow CH_3 CH{=}CH \cdot CH_2 OAc \quad 9\% \end{cases} \end{cases}$$

$$CH_3 \cdot CH{=}CH \cdot CH_3 \longrightarrow CH_3 \overset{OAc}{\underset{|}{CH}} \cdot \overset{PdOAc}{\underset{|}{CH}} \cdot CH_3 \longrightarrow CH_3 \cdot \overset{OAc}{\underset{|}{CH}} \cdot CH{=}CH_2 \quad {>}97\%$$

FIGURE 4.2

acetate are not in accord with observations on the oxidation of the butenes. The mechanism appears to be essentially the same as that discussed earlier for ketone formation. In the case of but-1-ene for example the intermediates, cf. Figure 4.2, are formed through reactions analogous to those in equations (i) to (v). The data available can be rationalised if it is assumed that Markovnikoff addition is preferred to non-Markovnikoff addition, and that elimination giving allylic acetate is preferred to enol acetate formation. The results obtained for the conversion of [2]H-propylene to propenyl (36%) and isopropenyl (64%) acetates[99] suggest that the elimination reaction is more complicated than would be envisaged. The enol acetates obtained from [2]H-propylene contain 75% of the original deuterium. Propylene reacts 2·8 times faster than [2]H-propylene but both give the same proportions of enol acetates. Thus, in the formation of isopropenyl acetate most of the deuterium has been retained and indicates that a deuterium migration must have occurred, the hydrogen atom lost coming from the methyl group, i.e.

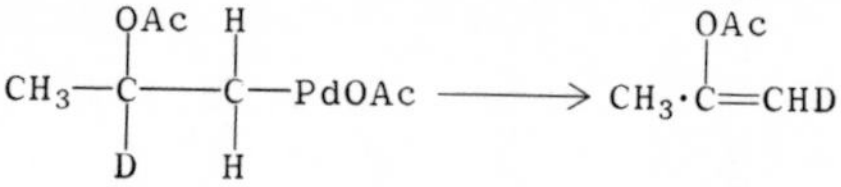

$$CH_3 - \overset{OAc}{\underset{D}{\overset{|}{C}}} - \overset{H}{\underset{H}{\overset{|}{C}}} - PdOAc \longrightarrow CH_3 \cdot \overset{OAc}{\underset{|}{C}}{=}CHD$$

However, there is no evidence for a similar process occurring in the formation of propenyl acetate.

The products obtained are greatly influenced by the solvent system used. The amount of crotyl acetate obtained from but-1-ene rises from 9% in acetic acid to 73% in a dimethylsulphoxide-acetic acid (7:3) mixture[61]. Amides, nitriles and sulphoxides are reported to increase the proportion of oxidation occurring at the terminal carbon atom of terminal olefins[14]. The addition of amides diminishes the amount of double bond migration in the olefin[15]. The proportion of allylic oxidation is increased by higher temperatures. At 150°C propylene gives mostly allyl acetate, while at 100° propenyl and isopropenyl acetates predominate[54]. Equilibration of allylic acetates is also observed, especially under prolonged reaction conditions[61].

The products of oxidation of aromatic hydrocarbons by palladium (II) are also influenced by reaction conditions. Under reaction conditions similar to those which convert olefins into vinyl esters benzene is oxidised in good yield ($\sim$ 70%) to diphenyl and small amounts of phenyl acetate[23, 104]. Addition of lithium acetate increases the proportion of phenyl acetate, whereas perchloric acid decreases it and also accelerates the reaction[23]. Similar oxidation of toluene in acetic acid yields mainly benzyl acetate, but addition of perchloric acid suppresses benzyl acetate formation in favour of the six isomeric nuclear dimerisation products. The isomer distribution is approximately $p \geqslant m \gg o$. The rates of oxidation of benzene, toluene, chlorobenzene and methyl benzoate to diphenyl derivatives are 0·088, 0·11, 0·07 and 0·04 mole^{-1}hr^{-1} respectively[104]. These values show that substituents have a weak polar effect with electron donating

CH$_3$ ring: 0·007, 0·028, 0·042   Cl ring: 0·0034, 0·019, 0·024   CO$_2$CH$_3$ ring: 0, 0·20, 0

groups increasing the reaction rate and electron attracting groups decreasing it. Analysis of the resulting reaction mixtures gives the isomer distribution shown in Table 4.4[104]. By combining the values for the rate constants and the reactivity ratios for p/m and o/m partial rate constants can be calculated.

TABLE 4.4 Oxidation of C_6H_5X[104]

X	$k_2$mole^{-1}hr^{-1}	Isomers of XC$_6$H$_4$·C$_6$H$_4$X				Reactivities per p, m and o position	
		4, 4′	3, 3′	3, 4′	3, 2′	p/m	o/m
H	0·088						
CH$_3$	0·11	20	20	35	25	1·5	0·25
Cl	0·07	15	25	40	20	1·28	0·18
C$_6$H$_5$O		mainly				$\gg$1	—
CO$_2$CH$_3$	0·04		mainly			<1	—

These partial rate constants are comparable to that observed in electrophilic substitution. The high proportion of meta isomers is characteristic of an electrophilic species with a fairly low specificity. The low amount of ortho isomer can be attributed to steric effects which are even more pronounced with di- and trisubstituted benzenes. o-Xylene gives (3) and (4) in a 2·7: 1 ratio and 1, 3-di-isopropylbenzene gives only (5). Mesitylene, p-di-t-butylbenzene and 1, 4-di-isopropylbenzene fail to undergo coupling.

(3)   (4)

(5)

The reactions of palladous acetate, platinous acetate, mercuric acetate and thallic acetate with benzene in acetic acid with perchloric acid have been compared[23]. The two former oxidants yield diphenyl and the latter phenylmercuric acetate and phenylthallium diacetate. The supposition that phenylpalladium and phenylplatinum derivatives are precursors of the diphenyl is further supported by the discovery that palladous acetate reacts with phenylboronic acid $(C_6H_5 . B(OH)_2)$ yielding diphenyl. The reaction of phenylboronic acid with mercury and thallium salts yields stable phenyl compounds as a result of metal-for-boron substitution reactions. The mode of breakdown of the phenyl palladium and platinum species is unclear. Oxidation of benzene to diphenyl by palladous sulphate in 20% sulphuric acid does not result in incorporation of deuterium into unreacted benzene when the reaction is carried out in deuterium oxide so that the inter-mediate is too short-lived to undergo acid cleavage. The generation of phenyl radicals seems unlikely in view of the insensitivity of the reaction to oxygen. As in the case of oxidation of olefins the initial stages of the reaction probably entail the formation of a benzene-palladium (II) complex (e.g. 6). This would accord with the ineffec-tiveness of palladous bromide and iodide as oxidants[104], which can be attributed to the greatly decreased coordinating power of palladi-um (II) in these compounds. Subsequent breakdown of (6) to phenyl acetate or phenylpalladium acetate is readily envisaged.

Other palladium catalysed conversions of olefins to vinyl derivatives have not been extensively studied. The reaction of ethylene with hydrogen chloride in the presence of palladous oxide and oxygen yields vinyl chloride[35], and a similar reaction with hydrogen cyanide gives acrylonitrile[36].

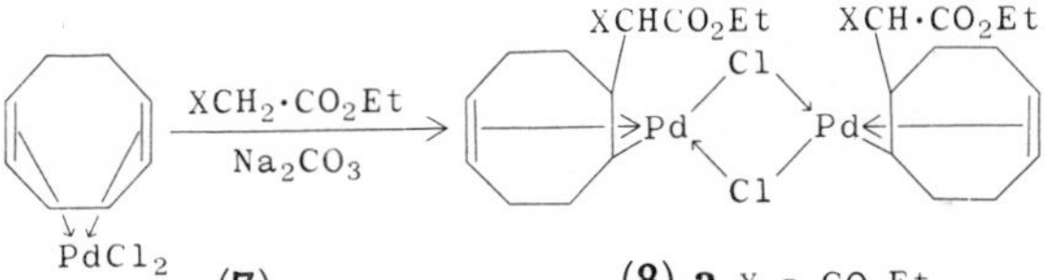

(7)　　(8) **a** X = CO_2Et

　　　　　b X = $CO\cdot CH_3$

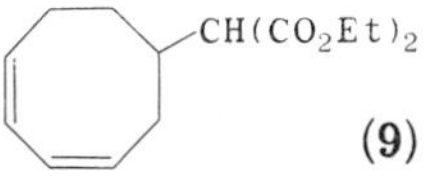

(9)

(8a)

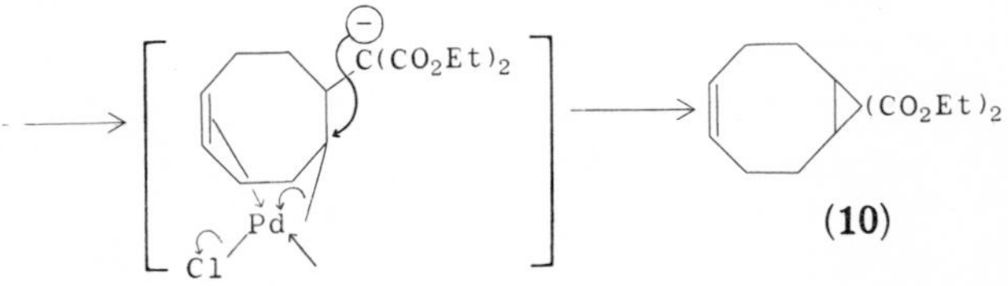

(10)

The reaction of olefin-palladium complexes with carbanions may proceed by a similar mechanism and promises to be of considerable synthetic utility. Reaction of the cycloocta-1, 5-diene-palladous chloride complex (7) with either diethyl malonate or ethyl aceto-acetate in the presence of sodium carbonate yields the complexes (8a) and (8b) respectively[102]. Treatment of (8a) with a weak base such as sodium carbonate or trimethylamine in boiling benzene gave (9) and palladium metal. The use of the strong base methylsulphinyl carbanion led by way of the malonate anion to (10). The reaction of π-allylpalladous chloride with diethyl malonate in the presence of sodium ethoxide gives a mixture of diethyl allylmalonate and diallylmalonate[103]. 2-Allylcyclohexanone is obtained following acid hydrolysis of the reaction mixture from the complex and 1-morpholinocyclohexene.

REFERENCES

1.　Anderson, C. B. and Winstein. S., J. Org. Chem., 1963, **28**, 605.

2.　Badische Anilin-und Soda-Fabrik, A. G., Belg. P. 658, 285; Chem. Abs., 1966, **64**, 6499.

3. Badische Anilin-und Soda Fabrik, A. G., Belg. P. 658, 801; Chem. Abs.. 1966, **64**, 6499.

4. Belov, A. P. and Moiseev, I. I., Izvest. Akad. Nauk S.S.S.R., Ser. khim, 1966, 139.

5. Belov, A.P., Moiseev, I.I. and Uvarova, N.G., Izvest. Akad. Nauk S.S.S.R., Ser. Khim., 1965, 2224.

6. Belov, A.P., Pek, G.Y. and Moiseev, I.I., Izvest. Akad Nauk S.S.S.R., Ser. Khim., 1965, 2204.

7. Berndt, W., Hoerning, L., Probst, O., Schmidt, W., Schwenk. U. and Weber, E., U.S.P. 3, 122, 586; Chem. Abs., 1964, **61**, 1758.

8. Berndt, W., Paszthory, E., Riemenschneider, W. and Weber, E., G.P. 1, 166, 174; Chem. Abs., 1964, **61**, 1759.

9. Bo, G., Achard, R., David, R. and Estienne, G., Fr.P. 1, 339, 614; Chem. Abs., 1964, **60**, 4012.

10. Bryant, D.R., McKeon, J. E. and Starcher, P. S., Fr.P. 1, 395, 129; Chem. Abs., 1965, **63**, 9816.

11. Bryant, D. R., McKeon, J. E. and Starcher, P. S., Belg.P. 646, 482; Chem. Abs., 1965, **63**, 17906.

12. Capp, C. W. and Harris, B. W., B.P. 1, 001, 539; Chem. Abs., 1965, **63**, 13083.

13. Chatt, J., Vallarino, L. M. and Venanzi, L. M., J. Chem. Soc., 1957, 3413.

14. Clark, D., Hayden, P., Walsh, W. D. and Jones, W. E., B.P. 964, 001; Chem. Abs., 1964, **61**, 13199.

15. Clark, D. and Walsh, W.D., B.P. 988, 011; Chem. Abs., 1965, **63**, 1706.

16. Clement, W. H. and Selwitz, C. M., Tetrahedron Letters, 1962, 1081.

17. Clement, W. H. and Selwitz, C. M., J. Org. Chem., 1964, **29**, 241.

18. Consortium für Elektrochemische Industrie, G. m. b. H., B.P. 884, 962; Chem. Abs., 1963, **59**, 5024.

19. Consortium für Elektrochemische Industrie, G.m.b.H., B.P. 884, 963; Chem. Abs., 1963, **59**, 5025.

20. Consortium für Elektrochemische Industrie, G.m.b.H., B.P. 892, 158; Chem. Abs., 1963, **59**, 13826.

21. Cotterill, C. B. and Dean, F., B.P. 941, 951; Chem. Abs., 1964, **60**, 4011.

22. Courtaulds Ltd., Neth. Appl. 6, 501. 821; Chem. Abs., 1966, **64**, 8042.

23. Davidson, J. M. and Triggs, C., Chem. and Ind., 1966, 457.

24. Dialer, K. and Riemenschneider, W., G.P. 1, 135, 441; Chem. Abs., 1963, **59**, 450.

25. Distillers Co. Ltd., Neth. Appl. 6, 409, 616; Chem. Abs., 1965, **63**, 8203

26. Dozono, T. and Shiba, T., Bull. Japan Petrol. Inst., 1963, **59**, 5829.

27. E. I. Du Pont de Nemours and Co., Neth. Appl. 6, 501, 904; 6, 501, 905; Chem. Abs., 1966, **64**, 4948.

28. Farbwerke Hoechst A.G., B.P. 898, 790; Chem. Abs., 1964, **61**, 8189.

29. Farbwerke Hoechst A.G., B.P. 922, 694; Chem. Abs., 1964, **61**, 6922.

30. Farbwerke Hoechst A.G., B.P. 938, 836; Chem. Abs., 1964, **60**, 11900.

31. Farbwerke Hoechst A.G., Belg. P. 626, 669; Chem. Abs., 1964, **60**, 9149.

32. Farbwerke Hoechst A.G., Belg. P. 635, 230; Chem. Abs., 1964, **61**, 11894.

33. Farbwerke Hoechst A.G., Fr.P. 1, 363, 747; Chem. Abs., 1964, **61**, 13196.

34. Farbwerke Hoechst A.G., Belg. P. 648, 304; Chem. Abs., 1965, **63**, 13082.

35. Fabwerke Hoechst A.G., Neth. Appl. 6, 504, 302; Chem. Abs., 1966, **64**, 12554.

36. Fabwerke Hoechst A.G., Neth. Appl. 6, 505, 608; Chem. Abs., 1966, **64**, 12557.

37. Hafner, W., Jira, R., Sedlmeier, J. and Smidt, J., Chem. Ber., 1962, **95**, 1575.

38. Hafner, W., Jira, R., Sedlmeier, J. and Smidt, J., G.P. 1, 137, 426; Chem. Abs., 1964, **60**, 2775.

38a. Hafner, W., Prigge, H. and Smidt, J., Annalen, 1966, **693**, 109.

39 Halcon International Inc., Neth. Appl. 6, 505, 941; Chem. Abs., 1966, **64**, 12551.

40. Harris, B. W., Fr.P. 1, 397, 054; Chem. Abs., 1965, **63**, 4164.

41. Harris, B. W., Fr.P. 1, 412, 151; Chem. Abs., 1966, **64**, 3362.

42. Henry, P. M., J. Amer. Chem. Soc., 1964, **86**, 3246.

43. Henry, P. M., J. Amer. Chem. Soc., 1966, **88**, 1595.

44. Hoernig, L., Paszthory, E., Riemenschneider, W. and Steinmetz, A., G.P. 1, 154, 450; Chem. Abs., 1964, **60**, 2775.

45. Holzrichter, H., Kroenig, W. and Frenz, B., Fr.P. 1, 346, 219; Chem. Abs., 1964, **60**, 11902.

46. Hüttel, R. and Christ, H., Chem. Ber., 1964, **97**, 1439.

47. Hüttel, R. and Christ, H., G.P. 1, 204, 648; Chem. Abs., 1966, **64**, 1964.

48. Hüttel, R., Kratzer, J. and Bechter, M., Chem. Ber., 1961, **94**, 766.

49. Imperial Chemical Industries Ltd., Belg.P. 628, 733; Chem. Abs., 1964, **60**, 15741.

50. Imperial Chemical Industries Ltd., Belg. P. 634, 595; Chem. Abs., 1964, **60**, 14394.

51. Imperial Chemical Industries Ltd., Belg. P. 635, 426; Chem. Abs., 1964, **61**, 11896.

52. Imperial Chemical Industries Ltd., Neth. Appl. 6, 412, 134; Chem. Abs., 1965, **63**, 13085.

53. Imperial Chemical Industries Ltd., Neth. Appl. 6, 413, 733; Chem. Abs., 1965, **63**, 17910.

54. Imperial Chemical Industries Ltd., Neth. Appl. 6, 501, 823; Chem. Abs., 1966, **64**, 4949.

55. Imperial Chemical Industries Ltd., Neth. Appl. 6, 501, 827; Chem. Abs., 1966, **64**, 4949.

56. Japan Oil Co. Ltd., B.P. 960, 195; Chem. Abs., 1964, **61**, 6922.

57. Japan Synthetic Chemical Ind. Co. Ltd., Belg. P. 618, 071; Chem. Abs., 1963, **59**, 12715.

58. Japan Synthetic Chemical Ind. Co. Ltd., Belg. P. 628, 848; Chem. Abs., 1964, **60**, 7917.

59. Jira, R., G. P. 1, 190, 931; Chem. Abs., 1965, **63**, 6865.

59a. Jira, R., Sedlmeier, J. and Smidt, J., Annalen, 1966, **693**, 99.

60. Joy, J. R. and Orchin, M., Z. anorg. Chem., 1960, **305**, 198.

61. Kitching, W., Rappoport, Z., Winstein, S. and Young, W. G., J. Amer. Chem. Soc., 1966, **88**, 2054.

62. Klass, D. L., U.S.P. 3, 147, 203; Chem. Abs., 1964, **61**, 13195.

63. Lum, D. W. and Koch, K., U.S.P. 3, 227, 747; Chem. Abs., 1966, **64**, 9599.

64. McKeon, J. E. and Starcher, P. S., Belg. P. 629, 885; Chem. Abs., 1964, **61**, 2976.

65. Matveev, K. I., Osipov, A. M., Odyakov, V. F., Suzdalnitskaya, Y. V., Bukhtoyarov, I. F. and Emelyanova, O. A., Kinetika i Kataliz, 1962, **3**, 661.

66. Matveev, K. I., Osipov, A. M. and Stroganova, L. N., U.S.S.R. 176, 258; Chem. Abs., 1966, **64**, 11086.

67. Moiseev, I. I., Belov, A. P. and Syrkin, Y. K., Izvest Akad. Nauk, S.S.S.R., Ser. khim., 1963, 1527.

68. Moiseev, I. I. and Vargaftik, M. N., U.S.S.R. 137, 508; Chem. Abs., 1962, **56**, 328.

69. Moiseev, I. I. and Vargaftik, M. N., Doklady Akad. Nauk S.S.S.R., 1966, **166**, 370.

70. Moiseev, I. I., Vargaftik, M. N. and Syrkin, Y. K., Doklady Akad. Nauk S.S.S.R., 1960, **130**, 820.

71. Moiseev, I. I., Vargaftik, M. N. and Syrkin, Y. K., Doklady Akad. Nauk S.S.S.R., 1963, **153**, 140.

72. Moiseev, I. I., Vargaftik, M. N. and Syrkin, Y. K., Doklady Akad. Nauk S.S.S.R., 1960, **133**, 377.

73. Moiseev, I. I., Vargaftik, M. N. and Syrkin, Y. K., Izvest. Akad. Nauk S.S.S.R., Ser.khim., 1963, 1144.

74. Neale, R. F., Fr.P. 1, 409, 190; Chem. Abs., 1966, **64**, 6567.

75. Nicolescu, I. V., Suceveanu, A. and Fordea, C., Rev. Roumaine Chim., 1965, **10**, 605.

76. Nikiforova, A. V., Moiseev, I. I. and Syrkin, Y. K., Zhur. obschei Khim., 1963, **33**, 3239.

77. Ninomiya, R., Sata, M. and Shiba, T., Bull. Japan Petrol. Inst., 1965, **7**, 31; Chem. Abs., 1965, **63**, 17829.

78. Omae, S. Matsushiro, K. and Yamamoto, H., Jap. P. 28, 090('65); Chem. Abs., 1966, **64**, 11089.

79. Pestrikov, S. V. and Kozik, B. L., Neftepererabotka i Neftekhim., Nauchn.-Tekhn.Sb., 1965, 35; Chem. Abs., 1965, **63**, 5521.

80. Phillips, F. C., Amer. Chem. J., 1894, **16**, 255.

81. Pullman, Inc., B.P. 1, 007, 815; Chem. Abs., 1966, **64**, 4949.

82. Riemenschneider, W., G.P. 1, 132, 910; Chem. Abs., 1963. **58**, 5517.

83. Riemenschneider, W., Schmidt, W. and Hoernig, L., G.P. 1, 136. 685; Chem. Abs., 1963, **58**, 12423.

84. Riemenschneider, W., Schwenk, U. and Weber, E., G. P. 1, 132, 553; Chem. Abs., 1962, **57**, 16407.

85. Schwenk, U. and Steinroetter, H., G. P. 1, 147, 569; Chem. Abs., 1963, **59**, 11260.

86. Shell Internationale Research Maatschappij N. V., Fr.P. 1, 385, 309; Chem. Abs., 1965, **63**, 8206.

87. Shell Internationale Research Maatschappij N. V., Neth. Appl. 287, 873; Chem. Abs., 1965, **63**, 9820.

88. Shell Internationale Research Maatschappij N.V., Neth. Appl. 6, 406, 180; Chem. Abs., 1966, **64**, 11091.

89. Smidt, J., Chem. and Ind., 1962, 54.

90. Smidt, J., Hafner, W. and Jira, R., U.S.P. 3, 080, 425; Chem. Abs., 1963, **59**, 7375.

91. Smidt, J., Hafner, W., Jira, R., Sedlmeier, J., Sieber, R., Rüttinger, R. and Kojer, H., Angew. Chem., 1959, **71**, 176.

92. Smidt, J., Hafner, W., Jira, R., Sieber, R., Sedlmeier, J., and Sabel, A., Angew. Chem., Internat. Edn., 1962, **1**, 80.

93. Smidt, J., Jira, R., Hafner, W. and Sedlmeier, J., G.P. 1, 142, 351; Chem. Abs., 1963, **59**, 7375.

94. Smidt, J., Jira, R., Sabel, A. and Sedlmeier, J., G.P. 1, 191, 362; Chem. Abs., 1965, **63**, 9820.

95. Smidt, J. and Krekeler, H., Erdöl u. Kohle, 1963, **16**, 560.

96. Smidt, J. and Sabel, A., G.P. 1, 127, 888; Chem. Abs., 1962, **57**, 3634.

97. Smidt, J. and Sieber, R., Angew. Chem., 1959, **71**, 626.

98. Steinmetz, A., Lenzmann, H., Probst. O. and Hoernig, L., G.P. 1, 145, 602; Chem. Abs., 1963, **59**, 8596.

99. Stern, E. W., Proc. Chem. Soc., 1963, 111.

100. Stern, E. W. and Spector, M. L., Proc. Chem. Soc., 1961, 370.

101. Teramoto, K., Oga, T., Kikuchi, S. and Ito, M., Yuki Gasei Kagaku Kyokai Shi, 1963, **21**, 298; Chem. Abs., 1963, **59**, 7339.

102. Tsuji, J. and Takahashi, H., J. Amer. Chem. Soc., 1965, **87**, 3275.

103. Tsuji, J., Takahashi, H. and Morikawa, M., Tetrahedron Letters, 1965, 4387.

104. Van Helden, R., and Verberg, G., Rec. Trav. chim., 1965, **84**, 1263.

105. Vargaftik, M. N., Moiseev, I. I. and Syrkin, Y. K., Doklady Akad. Nauk S.S.S.R., 1961, **139**, 1396.

106. Vargaftik, M. N., Moiseev, I. I. and Syrkin, Y. K., Doklady Akad. Nauk S.S.S.R., 1962, **147**, 399.

107. Vargaftik, M. N., Moiseev, I. I. and Syrkin, Y. K., Izvest. Akad. Nauk S.S.S.R., Ser. khim., 1963, 1147.

108. Vargaftik, M. N., Moiseev, I. I., Syrkin, Y. K. and Yakshin, V. V., Izvest. Akad. Nauk S.S.S.R., Ser. khim., 1962, 930.

109. Williamson, J. B., B.P. 1, 003, 347; Chem. Abs., 1965, **63**, 17909.

110. Yano, M., Mitsutani, A. and Tanaka, K., J. Chem. Soc. Japan, Ind. Chem. Sect., 1965, **68**, 1620.

111. Yoshioka, S., Ito, J., Omae, S., Matsushiro, K. and Yamamoto, H., Jap.P., 28, 089 ('65); Chem. Abs., 1966, **64**, 12553.

CHAPTER 5

THE MECHANISM OF THE CARBONYL INSERTION REACTION

As the mechanisms of organometallic reactions have gradually been clarified the existence of a general insertion reaction has become apparent[14]. Formally it can be defined as the addition of a covalent metal compound M-X to a neutral unsaturated molecule Y resulting in a product where the unsaturated molecule has inserted itself between the metal and the group which was originally bonded to the metal.

$$M\text{-}X + Y \longrightarrow M\text{-}Y\text{-}X$$

Two examples of this type of mechanism have already been encountered on p. 61 to p. 89. However, the most extensively investigated case is when the neutral unsaturated molecule is carbon monoxide. This as we shall see presently is inaptly referred to as the "carbonyl insertion reaction". As this reaction is fundamental to the ensuing four chapters it seems desirable to summarise in advance its basic mechanistic aspects.

Much of the present knowledge of this reaction has been derived from the study of alkyl and acyl manganese pentacarbonyls. Acetyl, phenylacetyl and benzoyl manganese pentacarbonyls are decarbonylated rapidly on heating to methyl, benzyl and phenyl manganese pentacarbonyls, respectively[8]. The reversibility of this process is demonstrated by the reconversion of methylmanganese pentacarbonyl to acetyl manganese pentacarbonyl by treatment with carbon monoxide at 35 atmospheres and room temperature.

$$R.Mn(CO)_5 + CO \rightleftharpoons R.CO.Mn(CO)_5$$

The rate of the carbonylation reaction decreases in the order $R = n$-propyl $>$ ethyl $>$ phenyl $>$ methyl $\gg$ benzyl, trifluoromethyl[4]. Studies using ^{14}CO have demonstrated that the acyl carbonyl group is derived from one of the carbon monoxide molecules initially coordinated to manganese and not from the entering carbon monoxide molecule[9] i.e.

$$CH_3Mn(CO)_5 + {}^{14}CO \longrightarrow CH_3COMn(CO)_4 \; {}^{14}CO$$

$$CH_3 \; {}^{14}CO.Mn(CO)_5 \longrightarrow CH_3Mn(CO)_4 \; {}^{14}CO + CO$$

112

The Carbonyl Insertion Reaction

The above finding suggests that other ligands can fulfil the role of the incoming carbon monoxide molecule in the carbonylation reaction. This has been verified with a range of ligands such as amines, phosphines, arsines, stibines and phosphites[1,6,16,17,19]. In accord with earlier comments on ease of carbonylation it has been noted that such reactions occur far less readily in the phenyl series than the methyl one[1,17] and not at all in the benzyl series[17]. Anions such as iodide, methoxide, thiocyanate and cyanide ions can also function as the incoming ligand[7].

$$CH_3Mn(CO)_5 + LiX \rightleftharpoons Li[CH_3COMn(CO)_4X]$$

The orientation of the entering ligand with respect to the acyl group depends on the nature of the ligand. With triphenylphosphine the *trans* isomer preponderates[18]. The *trans* to *cis* ratio is 4:1 at 5-10°C. and 7:3 at room temperature so that the ratio is also temperature dependent. Triphenylarsine gives almost exclusively the *trans* isomer while some amines form exclusively the *cis* isomer.

A number of kinetic studies have been made with the aim of determining whether the insertion of the carbonyl group and the coordination of another carbon monoxide molecule occur in two separate stages or are one concerted process. An initial study of the reaction of methyl manganese pentacarbonyl with carbon monoxide showed that the reaction was first order in each of the reactants, although the range of carbon monoxide concentration was low on account of its limited solubility[5]. A similar result was obtained for the reaction of methylmanganese pentacarbonyl with cyclohexylamine in hexane[19] But in tetrahydrofuran the reaction rate was independent of the concentration of amine. Analogous observations were made using other ethers or nitromethane as solvents and ligands such as triphenylphosphine and triphenylphosphite[6,19]. The simplest explanation of these results is that in non-coordinating solvents a concerted reaction occurs, but that in solvents such as tetrahydrofuran a solvent molecule initially enters the coordination sphere and is subsequently replaced.

The question as to whether the carbonylation is really an insertion of a carbon monoxide unit into the methylmanganese bond, or whether a migration of the methyl group from manganese to the carbonyl carbon occurs has been settled in favour of the latter pathway[20]. Since by the law of microscopic reversibility the decarbonylation reaction must follow the same pathway as the carbonylation reaction, only in reverse, the decarbonylation of *trans*-$CH_3CO . Mn(CO)_4Ph_3P$ has been studied. As can be seen from the accompanying diagrams the two mechanisms predict different sterochemistries for the resulting $CH_3 . Mn(CO)_4Ph_3P$. The product obtained has the *cis* stereochemistry showing that a methyl migration is involved.

Transition Metal Intermediates

Analogous carbonylation reactions have been observed with a range of well-defined transition metal compounds. These include methyl-π-cyclopentadienyliron dicarbonyl[8], methyl-π-cyclopentadienylmolybdenum tricarbonyl[8] and π-allylnickel halides[12]. Surprisingly alkylrhenium pentacarbonyls do not appear to undergo carbonylation although the converse decarbonylation of acylrhenium pentacarbonyls to the corresponding alkyl derivatives can be effected[15]. The carbonylation of alkylcobalt tetracarbonyls to acylcobalt tetracarbonyls is readily effected and the reaction is reversible. It has been estimated that acetylcobalt tetracarbonyl decarbonylates about 2250 times more rapidly than acetylmanganese pentacarbonyl[11].

Tertiary phosphine complexes of palladium and platinum alkyls and aryls $[MR_2(Et_3P)_2]$ have also been carbonylated[3], although in this case there are no initially coordinated carbon monoxide molecules. Carbonylation of *trans*-complexes such as (1) yields the *trans*-acyl complex (2). Again the reaction is reversible.

The palladium compounds are more readily carbonylated than the platinum ones and it has been suggested that, as these reactions presumably proceed through 5 or 6-coordinated intermediates, this is due to the greater ease with which palladium expends its coordination shell. A similar type of observation has been made with rhodium and iridium where although the complex $[CH_3RhICl(CO)(n\text{-}Bu_3P)]$ can be carbonylated the iridium analogue is unreactive under the same conditions[13].

Although no synthetic uses have as yet been reported several examples of sulphur dioxide insertion have been observed. Thus methyl or benzyl manganese pentacarbonyl reacts with sulphur dioxide forming the appropriate S-(alkylsulfinato)manganese penta-carbonyl[10].

$$R\text{-}Mn(CO)_5 \xrightarrow{SO_2} R.SO_2.Mn(CO)_5$$

Analogous reactions occur with methyl, ethyl or phenyl π-cyclopenta-dienyliron dicarbonyl[2]. As in the case of carbonylation reactions here also the phenyl compound reacts more slowly than the alkyl ones.

REFERENCES

1. Bannister, W. D., Greene, M. and Haszeldine, R. N., Chem. Comm., 1965, 54.

2. Bibler, J. P. and Wojcicki, A., J. Amer.Chem. Soc., 1964, 86, 5051.

3. Booth, G. and Chatt, J., Proc. Chem. Soc., 1961, 67.

4. Calderazzo, F. and Cotton, F. A., Proc. 7th Intern. Conf. Coordn. Chem. Stockholm, 1962, 296.

5. Calderazzo, F. and Cotton, F. A., Inorg. Chem., 1962, 1, 30.

6. Calderazzo, F. and Cotton, F. A., Chimica e Industria, 1964, 46, 1165.

7. Calderazzo, F. and Noack, K., J. Organometallic Chem., 1965, 4, 250.

8. Coffield, T. H., Kozikowski, J. and Closson, R. D., J. Org. Chem., 1957, 22, 598.

9. Coffield, T. H., Kozikowski, J. and Closson, R. D., Proc. of Int. Conf. on Coordn. Chem., Chem. Soc. Spec. Publn. No. 13, p. 126 (1959).

10. Hartman, F. A. and Wojcicki, A., J. Amer. Chem. Soc., 1966, 88, 844.

11. Heck, R. F., J. Amer. Chem. Soc., 1963, 85, 651.

12. Heck, R. F., J. Amer. Chem. Soc., 1963, 85, 2013.

13. Heck, R. F., J. Amer. Chem. Soc., 1964, 86, 2796.

14. Heck, R. F., Adv. Chem. Series, 1965, No. 49, p. 181.

15. Hieber, W., Braun, G. and Beck, W., Chem. Ber., 1960, 93, 901.

16. Keblys, K. A., U.S.P. 3, 081, 324. Chem. Abs., 1963, **59**, 6439.

17. Keblys, K. A. and Filbey, A. H., J. Amer. Chem. Soc., 1960, **82**, 4204.

18. Kraihanzel, C. S. and Maples, P. K., J. Amer. Chem. Soc., 1965, **87**, 5267.

19. Mawby, R. J., Basolo, F. and Pearson, R. G., J. Amer. Chem. Soc., 1964, **86**, 3994.

20. Mawby, R. J., Basolo, F. and Pearson, R. G., J. Amer. Chem. Soc., 1964, **86**, 5043.

CHAPTER 6

THE HYDROFORMYLATION REACTION

This reaction was first observed during investigations of the Fischer-Tropsch Process[169, 170]. Fundamentally, it entails conversion of an olefin into an aldehyde by reaction with synthesis gas, a mixture of carbon monoxide and hydrogen, in the presence of a cobalt catalyst.

$$R.CH = CH_2 + CO + H_2 \xrightarrow{Co_2(CO)_8} R.\overset{\displaystyle CHO}{\underset{\displaystyle |}{CH}}.CH_3 + R.CH_2CH_2.CHO$$

The reaction is also frequently known as the "Oxo" reaction and has acquired considerable industrial importance. As the reaction can be visualised as entailing the addition of the elements of formaldehyde (H-CHO) across the double bond the term "hydroformylation" was suggested[2]. Insofar as the products of such reactions are primarily aldehydes the term "hydroformylation" is clearly preferable.

Although, as will be seen later, other metal catalysts can be used the reaction is customarily performed with a cobalt catalyst. This is added either as preformed dicobalt octacarbonyl or formed in situ from a cobalt salt or cobalt metal. The temperatures normally employed are in the region 100-120°C, as at higher temperatures, 150-180°C, subsequent reduction of the aldehyde to a primary alcohol occurs. Carbon monoxide-hydrogen pressures of 200-300 atmospheres are typical.

The mechanism of the hydroformylation reaction has received much attention. The current mechanistic position is summarised in equations (1) to (7) where for simplicity only the formation of the straight chain isomer is depicted.

$$HCo(CO)_4 \rightleftharpoons HCo(CO)_3 + CO \tag{1}$$

$$\underset{\displaystyle CH_2}{\overset{\displaystyle RCH}{\|}} + HCo(CO)_3 \rightleftharpoons \underset{\displaystyle CH_2}{\overset{\displaystyle RCH}{\|}} \longrightarrow Co(CO)_3H \tag{2}$$

$$\underset{\displaystyle CH_2}{\overset{\displaystyle RCH}{\|}} \longrightarrow Co(CO)_3H + CO \rightleftharpoons RCH_2CH_2Co(CO)_4 \tag{3}$$

$$R\,CH_2CH_2Co(CO)_4 \rightleftharpoons R\,CH_2CH_2CO.Co(CO)_3 \tag{4}$$

$$R\,CH_2CH_2Co(CO)_3 + CO \rightleftharpoons R.CH_2CH_2CO.Co(CO)_4 \tag{5}$$

$$R. CH_2. CH_2 CO. Co(CO)_4 + HCo(CO)_4$$
$$\longrightarrow R CH_2CH_2CHO + Co_2(CO)_8 \quad (6)$$

$$R. CH_2CH_2CO Co(CO)_3 + H_2 \longrightarrow R CH_2CH_2CHO + H Co(CO)_3 (7)$$

It is now firmly established that cobalt hydrotetracarbonyl is the active species in the reaction, largely on the basis of its reactions with olefins at room temperature[21, 85, 91, 92, 117, 145, 158, 210, 211]. The predissociation of cobalt hydrotetracarbonyl into the hydrotricarbonyl species, equation (1), accounts for the inhibiting effect of carbon monoxide on the reaction of cobalt hydrotetracarbonyl with olefins[85], and is also in keeping with the mechanism of related substitution reactions[71]. The formation of the alkylcobalt tetracarbonyl compounds is well authenticated[71]. Similar complexes may be prepared from sodium tetracarbonylcobaltate and alkyl halides[68, 69, 70]. They have also been detected spectroscopically as transient intermediates in hydroformylation reactions[117]. The conversion of the alkylcobalt tetracarbonyl into the acylcobalt tetracarbonyl, equations (4) and (5), is the "carbonyl insertion" reaction reviewed in Chapter 5. Although it is formulated here as a stepwise process there is no reason to reject a concerted one. Initial claims for the detection of acylcobalt tricarbonyl as a stable species[21] have been withdrawn[66].

The final stage(s) of the hydroformylation reaction entails the reductive cleavage of the acylcobalt tetracarbonyl. While reduction of this intermediate by cobalt hydrotetracarbonyl equation (6) has been demonstrated it is doubtful whether this reaction would be fast enough to operate under normal hydroformylation conditions[21]. The alternative reduction by hydrogen has been effected at 25°C and 200 atmospheres and is probably the normal route. The inhibiting effect of carbon monoxide on this hydrogenation suggests that it is the acylcobalt tricarbonyl species which is reduced by hydrogen (equation (7)).

In view of the complexity of the reaction mechanism it is not surprising that the classical kinetic approach to its elucidation has been unsuccessful. The main value of the extensive kinetic studies that have been made is in defining the optimum reaction conditions[78, 124, 132, 133, 136, 138, 139, 158, 180]. Generally the reaction is characterised by a slow period while the cobalt catalyst is converted into cobalt hydrotetracarbonyl. It may be noted in passing that the conversion of dicobalt octacarbonyl is possibly a heterogeneous reaction[10, 11]. The rate of its formation increases with increasing hydrogen pressure and decreases with increasing carbon monoxide pressure[77, 81]. At constant catalyst concentration over the pressure range 120 to 380 atmospheres the hydroformylation reaction is first order with respect to olefin concentration and independent of total pressure. At pressures below 100 atmospheres the reaction is first order with respect to hydrogen. At constant hydrogen pressure the reaction rate is a maximum for partial carbon monoxide pressures of 8 to 12

The Hydroformylation Reaction

atmospheres. The rate is lower at higher carbon monoxide pressures and the reaction of carbon monoxide with dicobalt octacarbonyl[126] may contribute to this effect.

$$Co_2(CO)_8 + CO \rightleftharpoons Co_2(CO)_9$$

The reaction rate is increased by the addition of small amounts of organic bases, which accelerate the formation of cobalt carbonyls, and decreased by large amounts[82]. Pyridine and its homologues, with the exception of 2, 6-lutidine, are particularly effective[79, 83] whereas ammonia[102], primary and secondary amines[79, 83] are much less efficient. Tertiary amines almost inhibit the reaction. Variations in solvent medium are found to have little effect on the reaction rate for simple olefins whether paraffin hydrocarbons, aromatic hydrocarbons, ethers, ketones or alcohols are used as solvents. However, this is not so in all cases and it has been shown that for the hydroformylation of both acrylonitrile[88] and methyl acrylate[80] the order of decreasing solvent efficacy is alcohols > ketones > hydrocarbons.

Despite contrary claims[44, 59] it is clear that many sulphur containing compounds have a detrimental effect on the hydroformylation reaction, as a result of the formation of cobalt containing complexes devoid of catalytic activity[45, 95, 115, 116, 118, 119, 120]. The relative detrimental effects of elemental sulphur, carbon disulphide and ethanethiol are in the ratio 100:71:1·6.[100] Thiophens, their tetrahydro derivatives and sulphides have no effect. Small amounts of thiols, however, do reduce the induction period by accelerating cobalt carbonyl formation[130]. Other compounds which have adverse effects include oxygen[28], acetylenes[107], conjugated dienes[109] and α, β-unsaturated aldehydes[106]. The effects of the two latter classes of compounds are counteracted by alcohols.

The rate of hydroformylation of an olefin decreases with increasing alkyl substitution so that straight chain terminal olefins > straight chain internal olefins > branched chain olefins[207]. The effect is larger as the branching becomes closer to the double bond. These effects are readily intelligible in terms of the decreasing ease of π-complex formation, cf. equation (2). In cyclic olefins the reaction rate is at a minimum with cyclohexene so that the order of reactivity is cyclopentene > cyclohexene < cycloheptene > cyclooctene again in accord with ease of π-complex formation.

Despite the large number of compounds that have been hydroformylated in only two cases has the stereochemistry of addition been unambiguously established. The hydroformylation of 3, 4-di-0-acetyl-D-xylal (1)[172] and 3, 4, 6-tri-0-acetyl-D-glucal (2)[176] using deuterium in place of hydrogen demonstrates the occurrence of *cis* addition. The hydroformylation of Δ^5-unsaturated steroids[16, 17, 143], e.g. (3)

TABLE 6.1 Hydroformylation of Olefins

Olefin	Products	Yield, %	References
Ethylene	Propionaldehyde	72	2, 14, 15, 18, 32, 52, 53, 59, 62, 63, 138, 140, 141, 148, 158, 159, 164, 169, 189
Propylene	Butyraldehyde and *iso*butyraldehyde	80	22, 32, 51, 52, 53, 61, 62, 101, 103, 111, 125, 127, 130, 134, 140, 148, 153, 158, 159, 166, 169, 183, 189.
	n- and *iso*-butanol		26, 30, 49, 73, 90, 97, 114, 184.
But-1-ene	Valeraldehyde (70%) and 2-methylbutyraldehyde (30%)	56	93
	Pentan-1-ol and 2-methylbutanol		73, 90
But-2-ene	Pentan-1-ol and 2-methylbutanol		32, 90
Isobutylene	Isovaleraldehyde	90	13, 64, 65, 158, 159, 200
	3-Methylbutan-1-ol (47%), neopentyl alcohol (1%) isobutane	48	15, 90, 114, 201
Buta-1, 3-diene	*n*-Valeraldehyde and α-methylbutyraldehyde (1:1)	24	3, 129
	Hexanol, 2-methylpentanol, 2-ethylbutanol (5:4:1)	–	90
Pent-1-ene	C_6 aldehydes	80	38, 187

Pent-2-ene	C_6 aldehydes	75	1, 2
	Hexanol	62	26
2-Methylbut-1-ene	β-Methylvaleraldehyde	53	2
	4-Methylpentanol, 3-methylpentanol, 2, 3-dimethylbutanol (11:9:1)	–	90
3-Methylbut-1-ene	As above	–	90
2-Methylbut-2-ene	As above	–	90
Penta-1, 3-diene	Monoaldehydes	–	129
Penta-1, 4-diene	Pimelaldehyde, branched chain dialdehydes, cyclohexanecarboxaldehyde and monoaldehydes (after hydrogenation)	–	129
Isoprene	C_6 aldehydes	16	3
Hex-1-ene	Heptylaldehyde and 2-methylhexylaldehyde	64	2, 147, 157, 158
	Heptan-1-ol and 2-methylhexanol	46	90, 206
4-Methylpent-1-ene	2, 4-Dimethylvaleraldehyde, 2-ethyl-3-methyl-butyraldehyde, 3-methylhexaldehyde, 5-methyl-hexaldehyde	–	84
	5-Methylhexanol, 3-methylhexanol, 2, 4-di-methylpentanol (4:3:3)	–	90
2-Ethylbut-1-ene	β-Ethylvaleraldehyde	55	2
2, 3-Dimethylbut-1-ene	3, 4-Dimethylvaleraldehyde	45	3
	3, 4-Dimethylpentanol	–	90
2, 3-Dimethylbut-2-ene	3, 4-Dimethylpentanol	–	90
3, 3-Dimethylbut-1-ene	4, 4-Dimethylpentanol	–	90

(Continued overleaf)

TABLE 6.1 (continued)

Olefin	Products	Yield, %	References
1,2-Dimethylbuta-1,3-diene	C_7 aldehydes	22	3
2,3-Dimethylbuta-1,3-diene	3,4-Dimethylpentanal	45	3
Hexa-1,5-diene	Suberaldehyde, 2-methylheptanedial, monoaldehydes	-	129
Hept-1-ene	C_8 aldehydes		38, 55, 147
	Octanols	73	72, 73
2-Methylhexene and 2-ethylpentene	3-Methylheptan-1-ol and 3-ethylhexan-1-ol	-	195
Oct-1-ene	Nonaldehyde (27%), 2-methyloctaldehyde (31%)	58	36, 206
	C_9 alcohols	61	206
2,4,4-Trimethylpent-1-ene (di-isobutylene)	3,5,5-Trimethylhexanal	90	148
	3,5,5-Trimethylhexanol	60	58, 90, 114, 193
2-Ethylhexene (di-n-butene)	C_9 aldehydes	23	134
	3-Ethylheptanol	-	195
Myrcene	4,8-Dimethylnonanal	-	19, 29
Allocimene	4,8-Dimethylnonanal	70	19
Octadec-1-ene	C_{19} aldehydes	54	2, 131
Cyclopentene	Cyclopentanealdehyde	65	2
	Cyclopentylcarbinol	81	75, 90
Cyclopentadiene	Cyclopentanealdehyde	37	3

Substrate	Product	Yield	References
Cyclohexene	Cyclohexanealdehyde	80	32, 33, 53, 93, 134, 135, 156, 159
3-Vinylcyclohexene	Cyclohexylcarbinol	87	72, 75, 90, 114
	C_9 unsaturated aldehyde	–	32, 192
	C_{10} glycols and 3- and 4-ethylcyclohexyl-carbinols	–	34
Dipentene (limonene)	Aldehydes, alcohols and p-cymene	–	20, 29
α-Terpinene	Aldehydes, alcohols and p-cymene	–	29
Cycloheptatriene	Cycloheptylcarbinol	73	76
Cyclooctene	Cyclooctylcarbinol	84	75, 167
Cyclododecene	Cyclododecylcarbinol	90	75
cis, trans, trans-Cyclododeca-1, 5, 9-triene	Cyclododecylcarbinol	62	27
α-Pinene	2- or 3-Formyl-2, 6, 6-trimethylbicyclo[3, 1, 1]heptane	–	137, 156
	Methyl 3-(4'-methylcyclohexyl)butyl ether, 1-methyl-4-isopropylcyclohexanealdehyde, p-cymene	–	29
Camphene	Homoisocamphenilanaldehyde	68	98
Octalins	C_{11} aldehydes	32	2
Δ⁹-Octalin	No reaction	–	2
Dihydrodicyclopentadiene	Tricyclo[5, 2, 1, 0$^{2, 6}$]decane-4-methylol	90	182
	C_{11} aldehyde and C_{12} dialdehyde	–	181
	C_{11} alcohol and C_{12} diol	–	34

(Continued overleaf)

TABLE 6. 1 (continued)

Olefin	Products	Yield, %	References
3β, 20β-Dihydroxypregn-5-ene	6α-Hydroxymethylallopregnane-3β, 20β-diol	58	143
3β-Acetoxypregn-5-en-20-one	3β-Acetoxy-6α-hydroxymethyl-5α-pregnan-20-one	50	16, 17
11-Oxoprogesterone 3, 20 bis (ethyleneketal)	6-Hydroxymethylpregnan-3, 11, 20-trione 3, 20 bis(ethyleneketal)	–	16
Androst-5-ene-3β, 17β-diol diacetate	6-Hydroxymethyl-5α-androstane-3β-17β-diol diacetate	–	16
Sitosterol	6-Hydroxymethylsitostanol	–	16
Styrene	2-Phenylpropionaldehyde	30	1, 2, 156, 159
α-Methylstyrene	Aldehydes and alcohols (37%), isopropylbenzene (63%)	–	36, 179
1-Phenylbuta-1, 3-diene	Trace of n-butylbenzene and polymer	–	2, 3
Eugenol	3-Methoxy-4-hydroxy-1-propylbenzene (40%), 5, 6, 7, 8-tetrahydro-3-methoxy-2-naphthol (30-40%), 4-(3-methoxy-4-hydroxyphenyl)butan-1-ol (5%)	–	46
Isoeugenol	3-Methoxy-4-hydroxy-1-propylbenzene	80	46
1-Vinylnaphthalene	2-(1-Naphthyl)propionaldehyde	29	2
Vinylchloride	α-Chloropropionaldehyde	86	54, 99
1, 1-Dichloroethylene	α-Chloropropionaldehyde, α-chloroacraldehyde, 1, 1-dichloroethane	–	54
1, 2-Dichloroethylene	Traces of aldehyde	–	54

Allyl alcohol	γ-Hydroxybutyraldehyde	30	1, 2, 146, 156
Crotyl alcohol	3-Formylbutanol	–	156
Methyl vinyl ether	β-Methoxypropionaldehyde ?, its dimethylacetal and acetaldehyde	– / –	52
Butyl vinyl ether	α-Butoxypropionaldehyde	31	2
Allyl ethyl ether	β-Ethoxyisobutyraldehyde (30%), methylacrolein (6%), γ-ethoxybutyraldehyde (4%)	–	2 / 2
Allyl phenyl ether	γ-Phenoxybutyraldehyde	71	2, 122
Allyl 2-methylphenyl ether	γ-(2-Methylphenoxy)butyraldehyde	–	122
Allyl 4-fluorophenyl ether	γ-(4-Fluorophenoxy)butyraldehyde	–	121
Allyl 4-chlorophenyl ether	γ-(4-Chlorophenoxy)butyraldehyde	73	121
Allyl 2,4-dichlorophenyl ether	γ-(2,4-Dichlorophenoxy)butyraldehyde	92	121
Propylene oxide	β-Hydroxybutyraldehyde	–	212
Cyclohexene oxide	trans-2-Hydroxycyclohexanealdehyde dimer	40	171
Furan	2-Tetrahydrofurfuryl alcohol	35	32, 206
2,5-Dimethylfuran	2,5-Dimethyl-3-tetrahydrofurfuryl alcohol	23	206
2-Vinylfuran	Aldehydes formed but not isolated	–	2
5,6-Dihydro-4H-pyran	2-Hydroxymethyltetrahydropyran (78%), 3-hydroxymethyltetrahydropyran (8%)	– / –	41
6-Hydroxymethyl-5,6-dihydro-4H-pyran	2,6-Bishydroxymethyltetrahydropyran	77	41, 42
6,6-Bishydroxymethyl-5,6-dihydro-4H-pyran	1-Hydroxymethyl-6,8-dioxabicyclo[3,2,1]octane	78	41
2,6-Dimethyl-5,6-dihydro-4H-pyran	2,6-Dimethyl-3-hydroxymethyltetrahydropyran	43	41

(Continued overleaf)

126

TABLE 6. 1 (continued)

Olefin	Products	Yield, %	References
Methyl 6-methyl-5, 6-dihydro-4H-pyran-2-carboxylate	Methyl 6-methyltetrahydropyran-2-carboxylate and 6-methyl-3-hydroxymethyltetrahydropyran-2-carboxylic acid γ-lactone (1:1)	76	41
Dimethyl 6-methyl-5, 6-dihydro-4H-pyran-2, 3-dicarboxylate	Dimethyl 6-Methyltetrahydropyran-2, 3-dicarboxylate	77	41
3, 4-Di-0-acetyl-D-xylal	1, 5-Anhydro-4-deoxy-D-*arabino*-hexitol and 1, 5-anhydro-4-deoxy-L-*xylo*-hexitol	–	172, 174
	4, 5-Di-0-acetyl-2, 6-anhydro-3-deoxy-*aldehydo*-D-*lyxo*-hexose and -D-*xylo*-hexose	–	173
3, 4, 6-Tri-0-acetyl-D-glucal	4, 5, 7-Tri-0-acetyl-2, 6-anhydro-3-deoxy-D-*manno*-heptitol and -D-*gluco*-heptitol	–	176, 178
	Aldehydes of above sugars		173
3, 4, 6-Tri-0-acetyl-D-galactal	2, 6-Anhydro-3-deoxy-D-*galacto*-heptitol and -D-*lalo*-heptitol	–	175, 177, 178
Acraldehyde	Propionaldehyde	–	2
Acraldehyde diethyl acetal	Aldehydes formed but not isolated	–	2
4-Methyl-2-vinyl-*m*-dioxan	2-(2'-Formylethyl)-4-methyl-*m*-dioxan	28	56
Crotonaldehyde	Butyraldehyde	–	2
Crotonaldehyde diethyl acetal	Methylsuccindialdehyde mono diethyl acetal	50	156
Cyclohex-1-en-3-ylaldehyde	Cyclohexanedialdehyde	24	57

		Yield	References
Cyclohex-1-en-3-ylaldehyde diethyl acetal	Cyclohexanedialdehyde mono diethyl acetal	–	57
Methyl vinyl ketone	Methyl ethyl ketone	–	2
Mesityl oxide	Methyl isobutyl ketone	–	2
Vinyl acetate	α- and β-Acetoxypropionaldehydes	–	2, 5, 52
Allyl acetate	γ-Acetoxybutyraldehyde	75	1, 2
Methallyl acetate	γ-Acetoxy-β-methylbutyraldehyde	64	149
Allylidene diacetate	4,4-Diacetoxybutyraldehyde	75	2
Methyl acrylate	Methyl γ-oxobutyrate	54	32, 52, 53, 78, 79, 80, 198
	γ-Butyrolactone	69	39, 40
Ethyl acrylate	Ethyl γ-oxobutyrate	74	2, 60, 83, 144, 154
	γ-Butyrolactone	88	39, 40, 186
Methyl methacrylate	α-Methyl-γ-butyrolactone	51	39, 40
Methyl crotonate	δ-Valerolactone (72%) and β-methyl-γ-butyrolactone (20%)	–	39, 40, 186
Ethyl crotonate	Ethyl β-formylbutyrate and ethyl 5-oxopentanoate	71	1, 2, 60, 62, 154, 156, 160
	δ-Valerolactone (67%) and β-methyl-γ-butyrolactone (23%)	–	39, 186
Butyl crotonate	Butyl 5-oxopentanoate	37	60
Ethyl vinylacetate	δ-Valerolactone (52%) and β-methyl-γ-butyrolactone (17%)	–	39

(Continued overleaf)

TABLE 6.1 (continued)

Transition Metal Intermediates

Olefin	Products	Yield, %	References
Ethyl tiglate	α-Methyl-δ-valerolactone (31%), α-ethyl-γ-butyrolactone and α,β-dimethyl-γ-butyrolactone (21%)	–	39
Ethyl β,β-dimethylacrylate	β-Methyl-δ-valerolactone (88%) and β,β-dimethyl-γ-butyrolactone (1%)	–	39, 40, 186
Ethyl α,β,β-trimethylacrylate	α,β-Dimethyl-δ-valerolactone (55%) and α-isopropyl-γ-butyrolactone (31%)	–	39, 186
Ethyl α,α-dimethylvinylacetate	α,α-Dimethyl-δ-valerolactone (93%) and α,α,β-trimethyl-γ-butyrolactone (1%)	–	39, 40, 186
Ethyl sorbate	Ethyl diformylcaproates	49	60
	γ-Ethyl-δ-valerolactone (49%) and β-propyl-γ-butyrolactone (33%)	–	39, 186
Methyl cyclohex-3-enecarboxylate	Methyl formylcyclohexycarboxylate	–	32
Ethyl cyclohex-1-ene carboxylate	2-Hydroxymethylcyclohexanecarboxylic acid γ-lactone	23	39
Cyclohept-1-enylisobutyric acid	2-Hydroxymethylcycloheptylisobutyric acid lactone	81	39, 40, 186
Ethyl cyclohept-1-enylisobutyrate	2-Hydroxymethylcycloheptylisobutyric acid lactone	94	39, 186
Methyl undecylenate	C$_{13}$ aldehydes	71	1, 2
Methyl oleate	Mixed aldehyde esters	–	32, 131
Ethyl cinnamate	Ethyl β-phenylpropionate	–	2
	β-Phenyl-γ-butyrolactone	34	39, 40

Ethyl β-(2-furyl) acrylate	Ethyl β-(2-furyl) acrylate	–	2
Diethylmaleate	Diethyl α-formylsuccinate	62	60
	β-Carbethoxy-γ-butyrolactone	47	39, 40, 186
Diethyl fumarate	Diethyl α-formylsuccinate	48	1
	β-Carbethoxy-γ-butyrolactone	49	39, 40, 186
Diethyl itaconate	Diethyl α-(formylmethyl)succinate	56	60
Acrylonitrile	β-Cyanopropionaldehyde	57	2, 86, 87, 88, 89, 142
N-Acetylallylamine	N-Acetylaminobutyraldehyde	–	4
N,N-Diacetylallylamine	N,N-Diacetyl-γ-aminobutyraldehyde	87	4
N-Vinylphthalimide	N-Phthaloyl-α-amino and β-amino-propional-dehyde (3:7)	–	4
N-Phthaloylallylamine	N-Phthaloyl-γ-aminobutyraldehyde.	88	4
	N-Phthaloyl-β-aminoisobutyraldehyde and N-Phthaloyl-γ-aminobutyraldehyde (5:1)	–	4
N-Styrylphthalimide	β-Phthalimido-α-phenylpropionaldehyde	–	4
Vinyltrimethylsilane	α- and β-trimethylsilylpropionaldehydes (1:1)	16	24
Allyltrimethylsilane	Trimethylsilylbutyraldehyde	6	24

to (**4**), is also consistent with a *cis*-addition from the α-side although *trans* addition giving firstly the 5α, 6β-product, followed by epimerisation of the intermediary to the 5α, 6α-compound cannot be excluded.

The observed direction of addition to a double bond depends upon a number of factors. Any attempt at deductions from a perusal of the reported products obtained, (Table 6. 1) from hydroformylation reactions is of doubtful value. Many workers have failed to recognise that their products are mixtures of isomers. In some cases where relative amounts of isomers are recorded these may not be the original proportions as one of the isomers may have been selectively removed as the result of a side-reaction. One such reaction is the aldol condensation and another a Tischenko type oxidation-reduction between two molecules of aldehyde giving rise to an ester. The former reaction is probably partly responsible for the very low proportion of straight chain aldehyde sometimes observed[103]. The reported increase in the ratio of *n*- to *iso*-products by the addition of water may arise by partial inhibition of the aldol condensation[62]. These side-reactions and reduction of the aldehyde group can be minimised by adding alcohols or orthoformic esters so that the aldehydes are converted as formed into acetals, which are fairly stable under the reaction conditions[108].

Recent investigations have shown that the ratio of normal to iso aldehydes obtained from hydroformylation of terminal olefins depends on the reaction conditions. Hydroformylation of propylene at temperatures below 130°C shows that the total pressure has little effect on the ratio of n- to iso-butyraldehyde[22], but it is markedly increased by increasing the partial pressure of carbon monoxide[155, 161]. At higher temperatures, 130-200°C, the proportion of n-butyraldehyde increases with pressure[22, 101] up to 1, 000 atmospheres but falls off at still higher pressures. The interconversion of n- and iso-butyrylcobalt tetracarbonyls has been observed at room temperature[194]. In ether solution the two isomers are formed in almost equimolar amounts. Although carbon monoxide slows down the rate of isomerisation it does not alter the isomer proportions. Solvent has a marked effect on the isomerisation which is very slow in hexane or benzene. In dioxan only the iso isomer is isomerised and in ethyl acetate both isomers give a 7:3 n to iso ratio. A temperature effect occurs in the stoichiometric reaction of cobalt hydrotetracarbonyl with n-butyl vinyl ether, styrene and ethyl acrylate. Below 0°C these compounds are carbonylated on the α-carbon atom but at 25°C on the β-carbon atom[193]. In contrast the distribution of products from pent-1-ene was not affected by reaction temperature. These observations are readily rationalised as resulting from an equilibrium between the two isomeric acylcobalt tetracarbonyls arising from the reversibility of the stages indicated in equations (3), (4) and (5) on page 117.

$$
\begin{array}{l}
\text{CH}_2 \\
\| \\
\text{CH} \\
\\
\text{H} \cdot \text{C} \cdot \text{CH}_3 \\
\\
\text{CH}_2 \\
| \\
\text{CH}_3 \\
\textbf{(5)}
\end{array}
\left\{
\begin{array}{l}
\longrightarrow \quad \overset{\overset{\displaystyle \text{CH}_3}{|}}{\text{CH}_3 \cdot \text{CH}_2 \text{CH} \cdot \text{CH}_2 \cdot \text{CH}_2 \cdot \text{CHO}} \quad \textbf{(6)} \\
\\
\longrightarrow \quad (\text{CH}_3 \cdot \text{CH}_2)_2 \text{CH} \cdot \text{CH}_2 \text{CHO} \quad \textbf{(7)} \\
\\
\longrightarrow \quad \overset{\overset{\displaystyle \text{CH}_3 \quad \text{CH}_3}{| \quad\quad |}}{\text{CH}_3 \cdot \text{CH}_2 \cdot \text{CH} - \text{CH} \cdot \text{CHO}} \quad \textbf{(8)}
\end{array}
\right.
$$

The additional possibility of double bond migration must also be considered. This is particularly important with earlier work where Fischer-Tropsch catalysts, which necessitate higher reaction temperatures, have been used[12, 90, 169]. It is usually difficult to distinguish between the effects of double bond migration and changes in the direction of addition. An elegant approach to this problem based on the hydroformylation of (+) (S)-3-methylpent-1-ene (5) has clarified some of the more important factors governing double bond migration[163]. Conventional hydroformylation of (5) at 90°C gives (6), (7) and (8) in the ratio 93:4:3. Decreasing the carbon monoxide pressure below 5 atmospheres causes the ratio of (6) to (7) + (8) to decrease to 82:18. At 180°C the ratio of (6):(7):(8) becomes 61:20:19. The extent of racemisation of (6) parallels the formation of increasing amounts of (7) and (8) showing that the extent of double bond migration is small under conventional conditions. The hydroformylation of 4-methylpent-1-ene in the absence of a solvent shows the occurrence of extensive double bond migration[84]. However, using a solvent very little double bond isomerisation is observed unless the carbon monoxide pressure is greatly reduced[155]. Thus it is clear that availability of carbon monoxide in the liquid phase may have a marked effect on the reaction. A low availability of carbon monoxide may easily arise when a high rate of hydroformylation is accompanied by a too slow rate of carbon monoxide transfer from the gaseous to the liquid phase which prevents the saturation of the liquid medium.

As mentioned earlier the use of higher temperatures for the hydro formylation reaction can lead to reduction of the resulting aldehyde to a primary alcohol. Like the hydroformylation reaction the hydrogenation also involves cobalt hydrotetracarbonyl as demonstrated by stoichiometric reactions[47, 48, 209]. Kinetic studies[10, 113] show that the hydrogenation reaction is fastest at low partial pressures of carbon monoxide as is the hydroformylation reaction. However, carbon monoxide is more effective in suppressing the hydrogenation reaction and the rate is inversely proportional to the partial pressure of carbon monoxide. A reaction scheme analogous to that for the hydroformylation reaction has been proposed. This scheme provides an explanation for the origin of the formate esters which are frequently observed among the reaction products especially as the alcohols often give only minor amounts of formates under these conditions[123]. It may be noted that the complex, $(CH_3)_3C.O.CO.Co(CO)_4$ has been prepared directly from sodium tetracarbonylcobaltate and t-butylhypochlorite[67].

$$R.CH = 0$$
$$\downarrow$$
$$R.CHO + H.Co(CO)_3 \rightleftharpoons H.Co.(CO)_3$$

$$R.CHO \rightleftharpoons R.CH_2.O.Co(CO)_3$$
$$\downarrow$$
$$H Co(Co)_3$$

$$R.CH_2.O.Co(CO)_3 + H_2 \longrightarrow R.CH_2OH + H Co(CO)_3$$

$$R.CH_2.O.Co(CO)_3 + CO \rightleftharpoons R.CH_2O.Co(CO)_4 \rightleftharpoons$$
$$R.CH_2O.CO.Co(CO)_3.$$

$$R.CH_2O.CO.Co(CO)_3 + H_2 \longrightarrow R CH_2O.CO.H + H.Co(CO)_3$$

The observation that the addition of lead, mercury, bismuth, gold and zinc compounds inhibits aldehyde reduction[7, 9, 35, 104, 105] was first interpreted[8, 9] as implying a heterogenous reaction on metallic cobalt since these metals are known to poison solid catalysts. However, it has been shown that this is not the case[112] and that these compounds form cobalt carbonyl derivatives such as $Hg[Co(CO)_4]_2$ which are devoid of catalytic activity[105, 110]. The effect of these additives is to reduce the concentration of catalyst available. The effect on the aldehyde reduction is greater than on the hydroformylation reaction due to the slower rate of the former. Alcohol formation is claimed to be promoted by the addition of phosphines[26].

Hydroformylation of conjugated dienes usually gives saturated aldehydes[3]. This apparently results from reduction of intermediary $\alpha\beta$-unsaturated aldehydes which are readily reduced under these conditions[2, 206]. The reduction of $\alpha\beta$-unsaturated aldehydes may be circumvented by first converting them to acetals[56, 57, 156]. Furan, which often displays diene-like character, is converted to tetrahydrofurfuryl alcohol and 2, 5-dimethylfuran to 2, 5-dimethyl-3-tetrahydrofurfuryl alcohol[206]. Unsaturated acids, esters and nitriles are generally hydroformylated at the β-carbon atom, and, except in the case of ethyl cinnamate, reduction occurs to only a minor extent. If the hydroformylation of $\alpha\beta$-unsaturated esters is carried out at higher temperatures so that the aldehyde group is reduced good yields of lactones are obtained[39, 40, 186].

When the β-position of the double bond carries an alkyl group, as in methyl crotonate, the major product is the δ-lactone presumably resulting from an initial double bond migration.

Sorbic acid gives a mixture of γ- and δ-lactones the other double bond having been reduced. In the case of ethyl cinnamate the amount of hydrogenation can be greatly decreased by using a stoichiometric amount of catalyst.

The hydroformylation of acetylenes proceeds very sluggishly. Pent-1-yne gives only small amounts of n-hexanol and 2-methylpentanol while diphenylacetylene is reduced to stilbene[50].

Alcohols can also be hydroformylated providing rather higher temperatures, 160-180°C, are used, Table 6. 2. The same products are obtained as from the corresponding olefin. Thus $^{14}CH_3OH$ gives ethanol and acetic acid with all the carbon-14 at C. 2 and propionaldehyde with the carbon-14 equally distributed between C. 2 and C. 3[25]. Presumably the ethylcobalt tetracarbonyl intermediate undergoes isomerisation by way of equations (3) and (2), p. 117. Some hydrogenolysis of the alcohols occurs and is particularly marked in the case of benzyl alcohols where both the rates of reaction and extent of hydrogenolysis are increased by the introduction of nuclear *para* or *meta* electron releasing substituents[203].

TABLE 6.2 Hydroformylation of Alcohols

Alcohol	Products	References
Methanol	Acetaldehyde, propionaldehyde, methyl acetate	23
	39% Ethanol, 5% propanol, 1% butanol, 9% methyl acetate, 6.3% ethyl acetate, 8.5% methane	166, 202, 213
n-Propanol	n- and iso-Butyl alcohol, n- and iso-amyl alcohol	205, 213
Isopropanol	11% n- and iso-Butyl alcohols	205
t-Butanol	55% isovaleraldehyde, 10.5% pivalaldehyde	96, 128
	60% isoamyl alcohol, 4% neopentyl alcohol, 3% iso-butane, 3% isobutene	201, 205, 213
Cyclohexanol	Cyclohexylcarbinol	213
Pinacol	26% 3,4-dimethylpentanol, 17% pinacolone, 4% pinacolyl alcohol	208
Benzyl alcohol	31% 2-phenylethanol, 63% toluene	203, 204
p-Methylbenzyl alcohol	24% 2-(p-methylphenyl)-ethanol, 58% p-xylene	203
m-Methylbenzyl alcohol	36% 2-(m-methylphenyl)-ethanol, 52% m-xylene	203
p-t-Butylbenzyl alcohol	28% 2-(p-t-Butylphenyl)-ethanol, 54% p-t-butyltoluene	203
2,4,6-Trimethylbenzyl alcohol	18% 2-(2,4,6-trimethylphenyl)-ethanol, 58% 1,2,3,5-tetramethylbenzene	203

p-Hydroxymethylbenzyl alcohol	39% 2-(p-methylphenyl)-ethanol, 12% p-phenylene-β, β'-diethanol, 27% p-xylene	203
p-Methoxybenzyl alcohol	44% 2-(p-methoxyphenyl)-ethanol, 16% p-methoxy-toluene	203
m-Methoxybenzyl alcohol	2% 2-(m-methoxyphenyl)-ethanol, 23% m-methoxy-toluene	203
p-Chlorobenzyl alcohol	16% 2-(p-chlorophenyl)-ethanol, 41% p-chloro-toluene	203
m-Trifluromethylbenzyl alcohol	5% m-Methylbenzotrifluoride	203
p-Carbethoxybenzyl alcohol	27% ethyl (p-methyl)-benzoate	203
p-Nitrobenzyl alcohol	polymer of p-aminobenzyl alcohol	203

TABLE 6.3 Hydroformylation of Orthoformates

Orthoester	Product	Yield, %	References
Methyl orthoformate	Acetaldehyde dimethylacetal	90	150
Ethyl orthoformate	Propionaldehyde diethylacetal	83	150, 151, 162
n-Propyl orthoformate	n-Butyraldehyde di-n-propylacetal	80	150, 151
iso-Propylorthoformate	iso-Butyraldehyde di-isopropylacetal and small amounts of n-butyraldehyde	–	150, 151
n-Butyl orthoformate	Valeraldehyde di-n-butylacetal	81	150
iso-Butyl orthoformate	isoValeraldehyde and valeraldehyde (4:1)	–	150
(S)(−)-2-methylbutyl orthoformate	(S)(+)-3-Methylpentanal di-(S)(−)-2-methylbutyl acetal	50	150
(S)(+)-2-methylbutyl orthoformate	(+)-1,1-bis-(S)-2-methylbutoxy-3-methyl-pentane	95	151
2-Ethylbutyl orthoformate	3-Ethylpentanal di-2-ethylbutyl acetal	47	150
Benzyl orthoformate	Phenylacetaldehyde	–	150

Many of the disadvantages of using alcohols can be avoided by using their orthoformate esters[150, 151, 152, 162], Table 6.3. The overall reaction can be formulated as

$$2H. C(OR)_3 + CO + H_2 \longrightarrow RCH(OR)_2 + 2H. CO_2R + ROH$$

Particular advantages arise from the fact that little isomerisation occurs at the temperature used, so that *n*-propyl orthoformate gives only 1, 1-di-*n*-propoxybutane and the *iso*-propyl ester gives mostly the *iso*-butyraldehyde acetal with only minor amounts of the *n*-isomer. The absence of isomerisation often associated with hydroformylation of olefins is demonstrated by the conversion of (S) (-)-2-methylbutyl orthoformate into the (S) (-)-2-methylbutyl acetal of (S) (+)-3-methylpentanal[150]. Further, in contrast to the free alcohol, benzyl orthoformate gives the benzylacetal of phenylacetaldehyde. The use of the orthoesters of other acids has been investigated and the rates of reaction found to decrease in the order propionic > acetic > formic[152].

Epoxides can also be hydroformylated provided conditions are chosen so as to avoid their isomerisation, cf. p. 77. Ethylene oxide is converted to β-hydroxybutyraldehyde[212]. Cyclohexene oxide gives the dimer (9) of *trans*-2-hydroxy-cyclohexylaldehyde[171].

(9)

While cobalt carbonyls are normally used as catalysts in the hydroformylation reaction other metal carbonyls are also effective and are receiving increasing attention. Alkyl and acyl manganese pentacarbonyls are claimed in the patent literature[37, 94] to be better than cobalt carbonyls, whereas rhenium carbonyls are inactive[185]. Iron[52, 61] and nickel[52] compounds appear to be much less reactive than cobalt ones. Palladous chloride gives only a low yield of propionaldehyde from ethylene, due to extensive hydrogenation of ethylene[197].

Comparison of the efficacy of iridium, cobalt and rhodium catalysts for hydroformylation of olefins shows it to increase in this order[74]. Other investigations[6, 129, 187, 199] also indicate the high catalytic activity of rhodium compounds, which may be from a hundred to a thousand times more effective than cobalt catalysts[199]. The most effective rhodium compounds so far reported are 1, 2, 3-tris (triphenylphosphine)rhodium trichloride, 1, 2, 6-trispyridine-rhodium trichloride and the solvated rhodium (I) chlorine-bridged stannous chloride complex $[Rh_2Cl_2(SnCl_2, EtOH)_4]$.[147] These complexes catalyse the hydroformylation of hex-1-ene to *n*-heptaldehyde

(~70%) and 2-methylhexanoaldehyde (~20%) in ethanolic solution at 55°C and 90 atmospheres. At higher temperatures, 100°C, and higher proportions of hydrogen the aldehydes are reduced to alcohols.

In view of the sensitivity of cobalt carbonyl catalysed hydroformylation to reaction conditions it seems desirable to treat with reserve claims that rhodium catalysts produce larger amounts of branched chain aldehydes and also cause more double bond migration[199]. Differences in types of products obtained have been observed using different rhodium compounds. For example, hydroformylation of penta-1, 4-diene with rhodium oxide in ethanol yields after hydrogenation monoaldehydes, cyclohexanecarboxaldehyde, branched chain dialdehydes and pimelaldehyde while with rhodium trichloride the main products are cyclohexane and cyclohex-1-ene carboxaldehydes[129]. Dialdehydes are also formed by hexa-1, 5-diene but penta-1, 3-diene and buta-1, 3-diene yield only monoaldehydes. Hydrogenation of aldehydes to alcohols under hydroformylation conditions using rhodium trichloride only becomes important at high partial pressures of carbon monoxide[72].

Information on the use of ruthenium catalysts in the hydroformylation reaction is very limited and suggests that they are not very efficient[30, 188]. Relatively low conversions of olefins have been reported using ruthenium trichloride[188]. As with other catalysts hydrogenation of the aldehyde can occur[30]. The complex $[(Ph_3P)_4RuCl_2]$ gives only low yields of aldehydes because of the formation of the insoluble carbonyl complex $[(Ph_3P)_2RuCl_2(CO)_2]$.[38] The more soluble $[(Ph_3P)_2RuCl_3 . CH_3 . OH]$ gives higher yields of aldehydes from hept-1-ene. The most effective complex so far is $[(Ph_3P)_2Ru(CO)_3]$ which gives over 80% yields of aldehydes from pent-1-ene at 100°C and 100 atmospheres.

As mentioned earlier iron pentacarbonyl is not a satisfactory hydroformylation catalyst under the conditions usually employed but under alkaline conditions it converts olefins into alcohols or aldehydes[97, 168, 190]. At temperatures of 100-150°C and pressures of carbon monoxide in the 160 to 180 atmosphere region ethylene is converted to propanol and propionic acid[168], propylene to n- and iso-butanols[97, 165], but-1-ene to n- and iso-pentanols[168], and cyclopentene to cyclopentylcarbinol[191]. Under less strongly alkaline conditions the products are mainly the corresponding aldehydes[190,191]. The reaction medium is a very powerful reducing agent and converts benzaldehyde to benzyl alcohol[190,191], acetone to iso-propanol[191], allyl alcohol to propanol[168], methyl acrylate to methyl propionate[198], methyl vinyl ether to ethanol and methanol, and but-2-ene-1, 4-diol or dihydrofuran to butan-1, 4-diol[168]. These solutions also isomerise terminal olefins, hex-1-ene giving a mixture of hex-2-ene and hex-3-ene.

The Hydroformylation Reaction

REFERENCES

1. Adkins, H. and Krsek, G., J. Amer. Chem. Soc., 1948, **70**, 383.
2. Adkins, H. and Krsek, G., J. Amer. Chem. Soc., 1949, **71**, 3051.
3. Adkins, H. and Williams, J. L. R., J. Org. Chem., 1952, **17**, 980.
4. Ajinomoto Co., Inc., Fr.P. 1, 341, 874; Chem. Abs., 1964, **60**, 4061.
5. Ajinomoto Co., Inc., Fr.P. 1, 361, 797; Chem. Abs., 1964, **61**, 11894.
6. Alderson, T., U.S.P. 3, 020, 314; Chem. Abs., 1962, **56**, 9969.
7. Aldridge, C. L., U.S.P. 3, 091, 644; Chem. Abs., 1963, **59**, 11260.
8. Aldridge, C. L., Fasce, E. V. and Jonassen, H. B., J. Phys. Chem., 1958, **62**, 869.
9. Aldridge, C. L. and Jonassen, H. B., Nature, 1960, **188**, 404.
10. Aldridge, C. L. and Jonassen, H. B., J. Amer. Chem. Soc., 1963, **85**, 886.
11. Almasi, M. and Szabó, L., Acad. rep. populare Romine, Studii ceretărichim., 1960, **8**, 531.
12. Asahara, T., Yukagaku, 1956, 5, 19; Chem. Abs., 1960, **54**, 22319.
13. Badische Anilin-und Soda-Fabrik, B.P. 683, 267; Chem. Abs., 1954, **48**, 1422.
14. Badische Anilin-und Soda-Fabrik, B.P. 760, 409; Chem. Abs., 1957, **51**, 10562.
15. Barrick, P. L., U.S.P. 2, 542, 747; Chem. Abs., 1951, **46**, 7584.
16. Beal, P. F. and Rebenstorf, M. A., G.P. 1, 124, 942; Chem. Abs., 1963, **58**, 8021.
17. Beal, P. F., Rebenstorf, M. A. and Pike, J. E., J. Amer. Chem. Soc., 1959, **81**, 1231.
18. Bhattacharyya, S. K. and Subba Rao, B. C., J. Sci. Ind. Research, 1952, **11B**, 80.
19. Bordenca, C., U.S.P. 2, 790, 006; Chem. Abs., 1957, **51**, 14786.
20. Bordenca, C. and Lazier, W. A., U.S.P. 2, 584, 539; Chem. Abs., 1952, **46**, 8676.
21. Breslow, D. S. and Heck, R. F., Chem. and Ind., 1960, 467.
22. Brewis, S., J. Chem. Soc., 1964, 5014.
23. Brooks, R. E. U.S.P. 2, 457, 204; Chem. Abs., 1949, **43**, 3443.
24. Burkhard, C. A. and Hurd, D. T., J. Org. Chem., 1952, **17**, 1107; U.S.P. 2, 588, 083; Chem. Abs., 1952, **46**, 9120.

25. Burns, G. R., J. Amer. Chem. Soc., 1955, **77**, 6615.

26. Cannell, L. G., Slaugh, L. H. and Mullineaux, R. D., G.P. 1, 186, 455; Chem. Abs., 1965, **62**, 16054.

27. Chemische Werke Huels A.G., Fr.P. 1, 411, 448; Chem. Abs., 1966, **64**, 3380.

28. Čiha, M., Macho, V. and Štrešinka, J., Chem. Zvesti., 1959, **13**, 530.

29. Clement, W. H. and Orchin, M., Ind. and Eng. Chem. (Product Res. and Development), 1965, **4**, 283.

30. Cox, G. F. and Whitfield, G. H., B.P. 999, 461; Chem. Abs., 1965, **63**, 9811.

31. Crowe, B. F. and Elmer, O. C., U.S.P. 2, 742, 502; Chem. Abs., 1956, **50**, 16849.

32. E. I. du Pont de Nemours and Co., B.P. 614, 010; Chem. Abs., 1949, **43**, 4685.

33. Eisenmann, J. L. and Yamartino, R. L., G.P. 1, 146, 485; Chem. Abs., 1963, **59**, 11291.

34. Esso, B.P. 728, 913; Chem. Abs., 1956, **50**, 7852.

35. Esso, B.P. 864, 142; Chem. Abs., 1961, **55**, 18597.

36. Esso, Neth. Appl. 6, 400, 701; Chem. Abs., 1965, **62**, 5194.

37. Ethyl Corpn., B.P. 863, 277; Chem. Abs., 1962, **56**, 9969.

38. Evans, D., Osborn, J. A., Jardine, F. H. and Wilkinson, G., Nature, 1965, **208**, 1203.

39. Falbe, J., Huppes, N. and Korte, F., Chem. Ber., 1964, **97**, 863.

40. Falbe, J. and Korte, F., Angew. Chem. Internat. Edn., 1962, **1**, 657.

41. Falbe, J. and Korte, F., Chem. Ber., 1964, **97**, 1104.

42. Falbe, J. and Korte, F., U.S.P. 3, 159, 653; Chem. Abs., 1965, **62**, 9112.

43. Falbe, J. and Korte, F., Brennstoff-Chem., 1965, **46**, 276.

44. Field, E., U.S.P. 2, 683, 177; Chem. Abs., 1955, **49**, 10999.

45. Freund, M. and Marko, L., Brennstoff-Chem., 1965, **46**, 100.

46. Gaslini, F. and Nahum, L. Z., J. Org. Chem., 1964, **29**, 1177.

47. Goetz, R. W. and Orchin, M., J. Org. Chem., 1962, **27**, 3698.

48. Goetz, R. W. and Orchin, M., J. Amer. Chem. Soc., 1963, **85**, 2782.

49. Greene, C. R. and Meeker, R. E., Belg. P. 621, 833; Chem. Abs., 1963, **59**, 11259.

50. Greenfield, H., Wotiz, J. H. and Wender, I., J. Org. Chem., 1957, **22**, 542.

51. Gresham, W. F., B.P. 638, 754; Chem. Abs., 1950, **44**, 9473.

52. Gresham, W. F. and Brooks, R. E., U.S.P. 2, 497, 303; Chem. Abs., 1950, **44**, 4492.

53. Gresham, W. F., Brooks, R. E. and Bruner, W. M., U.S.P. 2, 549, 454; Chem. Abs., 1951, **44**, 8552.

54. Gut, G., El-Markhzangi, M. H. and Guyer, A., Helv. Chim. Acta, 1965, **48**, 1151.

55. Gwynn, B. H., U.S.P. 2, 748, 168; Chem. Abs., 1957, **51**, 1247.

56. Habeshaw, J. and Geach, C. J., B.P. 702, 206; Chem. Abs., 1955, **49**, 5514.

57. Habeshaw, J. and Rae, R. W., B.P. 702, 201; Chem. Abs., 1955, **49**, 5514.

58. Habeshaw, J. and Thornes, L. S., B.P. 702, 195; Chem. Abs., 1955, **49**, 5513.

59. Hagemeyer, H. J., U.S.P. 2, 691, 045; Chem. Abs., 1955, **49**, 14797.

60. Hagemeyer, H. J. and Hull, D. C., U.S.P. 2, 610, 203; Chem. Abs., 1953, **47**, 5960.

61. Hagemeyer, H. J. and Hull, D. C., U.S.P. 2, 694, 734; Chem. Abs., 1955, **49**, 15947.

62. Hagemeyer, H. J. and Hull, D. C., U.S.P. 2, 694, 735; Chem. Abs., 1955, **49**, 15947.

63. Hasek, R. H., U.S.P. 2. 691, 046; Chem. Abs., 1955, **49**, 14797.

64. Haubner, H. and Hagen, W., G.P. 945, 685; Chem. Abs., 1958, **52**, 16202.

65. Haubner, H. and Hagen, W., G.P. 964, 857; Chem. Abs., 1959, **53**, 14005.

66. Heck, R. F., J. Amer. Chem. Soc., 1963, **85**, 1220.

67. Heck, R. F., J. Organometallic Chem., 1964, **2**, 195.

68. Heck, R. F. and Breslow, D. S., J. Amer. Chem. Soc., 1960, **82**, 750.

69. Heck, R. F. and Breslow, D. S., J. Amer. Chem. Soc., 1960, **82**, 4438.

70. Heck, R. F. and Breslow, D. S., J. Amer. Chem. Soc., 1961, **83**, 1097.

71. Heck, R. F. and Breslow, D. S., J. Amer. Chem. Soc., 1961, **83**, 4023.

72. Heil, B. and Marko, L., Chem. Ber., 1966, **99**, 1086.

73. Hughes, V. L. and Kirshenbaum, I., Ind. Eng. Chem., 1957, **49**, 1999.

74. Imyanitov, N. S. and Rudkovskii, D. M., Neftekhimiya, 1963, **3**, 198.

75. Inventa A-G. für Forschung und Patentsverwertung, Fr. P. 1, 404, 182; Chem. Abs., 1965, **63**, 16230.

76. Inventa A-G. für Forschung und Patentsverwertung, B. P. 1, 007, 627; Chem. Abs., 1966, **64**, 4968.

77. Iwanaga, R., Bull. Chem. Soc. Japan, 1962, **35**, 774.

78. Iwanaga, R., Bull. Chem. Soc. Japan, 1962, **35**, 778.

79. Iwanaga, R., Bull. Chem. Soc. Japan, 1962, **35**, 865.

80. Iwanaga, R., Bull. Chem. Soc. Japan, 1962, **35**, 869.

81. Iwanaga, R., Fujii, T., Wakamatsu, H., Yoshida, T. and Kato, J., J. Chem. Soc. Japan, Ind. Chem. Sect., 1960, **63**, 960.

82. Iwanaga, R., Fujii, T., Wakamatsu, H., Yoshida, T. and Kato, J., J. Chem. Soc. Japan, Ind. Chem. Sect., 1960, **63**, 1754.

83. Iwanaga, R., Mori, Y. and Yoshida, T., Jap. P. 8177('57); Chem. Abs., 1958, **52**, 14661.

84. Johnson, M., Chem. and Ind., 1963, 684; J. Chem. Soc., 1963, 4859.

85. Karopinka, G. L. and Orchin, M., J. Org. Chem., 1961, **26**, 4187.

86. Kato, J., Ito, T., Yabe, Y., Iwanaga, R. and Yoshida, T., J. Chem. Soc. Japan, Ind. Chem. Sect., 1963, **58**, 4420.

87. Kato, J., Wakamatsu, H. and Ishihara, H., Jap. P., 2, 574('61); Chem. Abs., 1962, **56**, 9977.

88. Kato, J., Wakamatsu, H., Iwanaga, R. and Yoshida, T., J. Chem. Soc. Japan, Ind. Chem. Sect., 1961, **64**, 2139.

89. Kato, J., Wakamatsu, H., Komatsu, T., Iwanaga, R. and Yoshida, T., J. Chem. Soc. Japan, Ind. Chem. Sect., 1961, **64**, 2142.

90. Keulemans, A. I. M., Kwantes, R. and Bavel, Th. van, Rec. Trav. chim., 1948, **67**, 298.

91. Kirch, L. and Orchin, M., J. Amer. Chem. Soc., 1958, **80**, 4428.

92. Kirch, L. and Orchin, M., J. Amer. Chem. Soc., 1959, **81**, 3597.

93. Klemchuk, P., G.P., 1, 165, 568; Chem. Abs., 1964, **61**, 2988.

94. Klopfer, O. E. H., U.S.P., 3, 050, 562; Chem. Abs., 1962, 57, 13217.

95. Klumpp, E., Marko, L. and Bor, G., Chem. Ber., 1964, 97, 926.

96. Kröper, H., Häuber, H. and Hagen, W., G.P. 821, 936; Chem. Abs., 1959, 53, 222.

97. Kutepow, N. and Kindler, H., Angew. Chem., 1960, 72, 802.

98. LoCicero, J. C. and Johnson, R. T., J. Amer. Chem. Soc., 1952, 74, 2094.

99. Lonza Ltd., Fr.P. 1, 397, 779; Chem. Abs., 1965, 63, 6864.

100. Macho, V., Chem. Zvesti, 1961, 15, 181.

101. Macho, V., Chem. prümysl, 1961, 11, 630.

102. Macho, V., Chem. Zvesti, 1962, 16, 73.

103. Macho, V., Chem. prümysl, 1962, 12, 240.

104. Macho, V., Chem. Zvesti, 1963, 17, 525.

105. Macho, V., Ropa Uhlie, 1964, 6, 297; Chem. Abs., 1965, 62, 8996.

106. Macho, V., Chem. Zvesti, 1964, 18, 890.

107. Macho, V., Marko, M. and Čiha, M., Chem. Zvesti., 1961, 15, 830.

108. Macho, V., Marko, M. and Čiha, M., Chem. Zvesti, 1962, 16, 65.

109. Macho, V., Marko, M., Čiha, M. and Štrešinka, J., Czech. P. 104, 691; Chem. Abs., 1963, 59, 11261a.

110. Macho, V., Mistrik, E. J. and Čiha, M., Coll. Czech. Chem. Comm., 1964, 29, 826.

111. Macho, V., Mistrik, E. J. and Štrešinka, J., Chem. prümysl, 1963, 13, 343.

112. Markó, L., Khim. i Tekhnol. Topliv i Mosel, 1960, 5, 19; Chem. Abs., 1961, 55, 2075.

113. Markó, L., Proc. Chem. Soc., 1962, 67.

114. Markó, L., Chem. and Ind., 1962, 260.

115. Markó, L., Bor. G. and Almásy, G., Chem. Ber., 1961, 94, 847.

116. Markó, L., Bor, G., Almásy, G. and Klumpp, E., Magy Asvanyolaj Foldgaz Kiserl. Int. Kozlemeny, 1962, 3, 242; Chem. Abs., 1963, 58, 4148.

117. Markó, L., Bor, G., Almásy, G. and Szabo, P., Brennstoff-Chem., 1963, 44, 184.

118. Markó, L., Bor, G. and Klumpp, E., Chem. and Ind., 1961, 1491.

119. Markó, L., Bor. G. and Klumpp, E., Angew. Chem. Internat. Edn., 1963, **2**, 210.

120. Markó, L., Bor, G., Klumpp, E., Markó, B. and Almásy, G., Chem. Ber., 1963, **96**, 955.

121. Markó, V., Strešinka, J., Macho, V. and Gregor, F., Czech. P. 109, 561; Chem. Abs., 1964, **60**, 13188.

122. Markó, V., Strešinka, J., Macho, V. and Gregor, F., Czech. P. 109, 618; Chem. Abs., 1964, **60**, 13188.

123. Markó, L., and Szabo, P., Chem. Tech. (Berlin), 1961, **13**. 482.

124. Martin, A. R., Chem. and Ind., 1954, 1536.

125. Matsuda, A. and Uchida, H., Tokyo Kogyo Shikensho Hokoku, 1962, **57**, 50; Chem. Abs., 1965, **62**, 7625.

126. Metlin, S., Wender, I. and Sternberg, H. W., Nature, 1959, **183**, 457.

127. Millidge, A. E., Fr. P. 1, 411, 602; Chem. Abs., 1966, **64**, 598.

128. Mönkemeyer, K., U.S.P. 2, 770, 655; Chem. Abs., 1957, **51**, 5817.

129. Morikawa, M., Bull. Chem. Soc. Japan, 1964, **37**, 379.

130. Morikawa, M., Bull. Chem. Soc. Japan, 1964, **37**, 430.

131. Natta, G. and Beati, E., B.P. 646, 424; Chem. Abs., 1951, **45**, 5714.

132. Natta, G. and Ercoli, R., Chimica e Industria, 1952, **34**, 503.

133. Natta, G., Ercoli, R. and Castellano, S., Chimica e Industria, 1955, **37**, 6.

134. Natta, G., Ercoli, R. and Castellano, S., Ital. P. 516, 716; Chem. Abs., 1958, **52**, 1221.

135. Natta, G., Ercoli, R. and Castellano, S., U.S.P. 3, 008, 996; Chem. Abs., 1962, **57**, 9694.

136. Natta, G., Ercoli, R., Castellano, S. and Barbieri, F.H., J. Amer. Chem. Soc., 1954, **76**, 4049.

137. Natta, G. and Pino, P., Chimica e Industria, 1949, **31**, 109.

138. Natta, G., Pino, P. and Beati, E., Chimica e Industria, 1949, **31**, 111.

139. Natta, G., Pino, P. and Beati, E., Chimie et Industrie, 1950, **63**, 464.

140. Niwa, A., Kikuchi, Y., Kamimura, S. and Onishi, M., Jap. P. 1107('57); Chem. Abs., 1959, **52**, 4680.

141. Niwa, M., Kikuchi, Y., Kamimura, S. and Onishi, M., U.S.P. 2, 992, 275; Chem. Abs., 1962, **56**, 2333.

142. Noguchi Research Foundation, Fr.P. 1, 370, 004; C.A. 1965, **63**, 9823.

143. Nussbaum, A. L., Popper, T. L., Oliveto, E. P., Friedman, S. and Wender, I., J. Amer. Chem. Soc., 1959, **81**, 1228.

144. Ohashi, K. and Suzuki, S., J. Chem. Soc. Japan, Ind. Chem. Sect., 1953, **56**, 792.

145. Orchin, M., Kirch, L. and Goldfarb, I., J. Amer. Chem. Soc., 1956, **78**, 5450.

146. Orchin, M. and Wender, I, "Catalysis", Reinhold Publ. Corp., New York, N.Y., 1957, vol. V, p. 1.

147. Osborn, J. A., Wilkinson, G. and Young, J. F., Chem. Comm., 1965, 17.

148. Osumi, Y., Yamaguchi, M., Onoda, T. and Onishi, M., Jap. P. 22, 735('65); Chem. Abs., 1966, **64**, 4943.

149. Parham, W. E. and Holmquist, H. E., J. Amer. Chem. Soc., 1951, **73**, 913.

150. Piacenti, F., Gazzetta, 1962, **92**, 225.

151. Piacenti, F., Ciono, C. and Pino, P., Chem. and Ind., 1960, 1240.

152. Piacenti, F. and Neggiani, P. P., Chimica e Industria, 1962, **44**, 1396.

153. Piacenti, F. and Pino, P., Belg.P. 613, 606; Chem. Abs., 1963, **58**, 451c.

154. Piacenti, F., Pino, P. and Bertolaccini, P.L., Chimica e Industria, 1962, **44**, 600.

155. Piacenti, F., Pino, P., Lazzaroni, R. and Bianchi, M., J. Chem. Soc. (C), 1966, 488.

156. Pino, P., Gazzetta, 1951, **81**, 625.

157. Pino, P. and Ercoli, R., Ricerca sci., 1953, **23**, 1231.

158. Pino, P., Ercoli, R. and Calderazzo, F., Chimica e Industria, 1955, **37**, 782.

159. Pino, P. and Paleari, C., Gazzetta, 1951, **81**, 646.

160. Pino, P., Piacenti, F. and Mantica, E., Chimica e Industria, 1956, **38**, 34.

161. Pino, P., Piacenti, F. and Neggiani, P. P., Chem. and Ind., 1961, 1400.

162. Pino, P., Piacenti, F. and Neggiani, P. P., Chimica e Industria, 1962, **44**, 1367.

163. Pino, P., Pucci, S. and Piacenti, F., Chem. and Ind., 1963, 294.

164. Pistor, H. J., Kölsch, W., and Eckert, E., G. P. 921, 935; Chem. Abs., 1959, **53**, 223.

165. Reed, H. W. B. and Lenel, P. O., B. P. 794, 067; Chem. Abs., 1959, **53**, 218.

166. Reppe, W. and Friederich, H., G. P. 894, 403; Chem. Abs., 1956, **50**, 16830.

167. Reppe, W., Schlichting, O., Klager, K. and Toepel, T., Annalen, 1948, **560**, 1.

168. Reppe, W., and Vetter, H., Annalen, 1953, **582**, 133.

169. Roelen, O., U.S.P., 2, 327, 066; Chem. Abs., 1944, **38**, 550.

170. Roelen, O., Angew. Chem., 1948, **A60**, 62.

171. Roos, L., Goetz, R. W. and Orchin, M., J. Org. Chem., 1965, **30**, 3023.

172. Rosenthal, A. and Abson, D., Canad. J. Chem., 1964, **42**, 1811.

173. Rosenthal, A. and Abson, D., J. Amer. Chem. Soc., 1964, **86**, 5356.

174. Rosenthal, A. and Abson, D., Canad. J. Chem., 1965, **43**, 1318.

175. Rosenthal, A. and Abson, D., Canad. J. Chem., 1965, **43**, 1985.

176. Rosenthal, A. and Koch, H. J., Canad. J. Chem., 1965, **43**, 1375.

177. Rosenthal, A. and Read, D., Canad. J. Chem., 1957, **35**, 788.

178. Rosenthal, A. and Read, D., Methods in Carbohydrate Chemistry, 1963, **2**, 450.

179. Rudkovskii, D. M. and Imyanitov, N. S., Zhur. priklad. Khim., 1962, **35**, 2719.

180. Rudkovskii, D. M., Trifel, A. G. and Alekseeva, K. A., Khim. Prom., 1959, 652.

181. Ruhr-Chemie A. G., B.P. 750, 144; 765, 742; Chem. Abs., 1957, **51**, 9680, 12970.

182. Ruhr-Chemie A. G., B.P. 779, 241; Chem. Abs., 1958, **52**, 1224.

183. Schiller, G., G.P. 953, 605; Chem. Abs., 1959, **53**, 11226.

184. Schreyer, R. C., U.S.P. 2, 564, 130; Chem. Abs., 1953, **47**, 142.

185. Selin, T. G., U.S. Dept. Comm., Office Tech. Serv., P.B. Rept. 133796 (1960); Chem. Abs., 1962, **56**, 4142.

186. Shell International Research Maatschappij N.V., Belg.P. 616, 141; Chem. Abs., 1963, **59**, 1495.

187. Slaugh, L. H. and Mullineaux, R. D., U.S.P. 3, 239, 566; Chem. Abs., 1966, **64**, 15745.

188. Smith, P. and Jaeger, H. H., Ger.P. 1, 159, 926; Chem. Abs., 1964, **60**, 14389.

189. Staib, J. H., Guyer, W. R. and Slotterbeck, O. C., U.S.P. 2, 864, 864. Chem. Abs., 1959, **53**, 9063.

190. Sternberg, H. W., Markby, R. and Wender, I., J. Amer. Chem. Soc., 1956, **78**, 5704.

191. Sternberg, H. W., Markby, R. and Wender, I., J. Amer. Chem. Soc., 1957, **79**, 6116.

192. Stewart, J., Staib, J. H. and Knoth, F., U.S.P. 2, 810, 748; Chem. Abs., 1958; **52**, 3857.

193. Takegami, Y., Yokokawa, C., Watanabe, Y., Masada, H. and Okuda, Y., Bull. Chem. Soc. Japan, 1965, **37**, 1190.

194. Takegami, Y., Yokokawa, C., Watanabe, Y., Masada, H. and Okuda, Y., Bull. Chem. Soc. Japan, 1965, **38**, 787.

195. Taylor, A. W. C., B.P. 798, 541; Chem. Abs., 1959, **53**, 2089.

196. Taylor, A. W. C. and Lamb, S. A., B.P. 684, 673; Chem. Abs., 1954, **48**, 1421.

197. Tsuji, J., Iwamoto, N. and Morikawa, M., Bull. Chem. Soc. Japan, 1965, **38**, 2213.

198. Uchida, H. and Bando, K., Bull. Chem. Soc. Japan, 1956, **29**, 953.

199. Wakamatsu, H., J. Chem. Soc. Japan, 1964, **85**, 227.

200. Warren, G. W., Haskin, J. F., Kouney, R. E. and Yarborough, V. A., Analyt. Chem., 1959, **31**, 1624.

201. Wender, I., Feldman, J., Metlin, S., Gwynn, B.H. and Orchin, M., J. Amer. Chem. Soc., 1955, **77**, 5760.

202. Wender, I., Friedel, R. A. and Orchin, M., Science, 1951, **113**, 206.

203. Wender, I., Greenfield, H., Metlin, S. and Orchin, M., J. Amer. Chem. Soc., 1952, **74**, 4079.

204. Wender, I., Greenfield, H. and Orchin, M., J. Amer. Chem. Soc., 1951, **73**, 2656.

205. Wender, I., Levine, R. and Orchin, M., J. Amer. Chem. Soc., 1949, **71**, 4160.

206. Wender, I., Levine, R. and Orchin, M., J. Amer. Chem. Soc., 1950, **72**, 4375.

207. Wender, I., Metlin, S., Ergun, S., Sternberg, H. W. and Greenfield, H., J. Amer. Chem. Soc., 1956, **78**, 5401.

208. Wender, I., Metlin, S. and Orchin, M., J. Amer. Chem. Soc., 1951, **73**, 5704.

209. Wender, I., Orchin, M. and Storch, H. H., J. Amer. Chem. Soc., 1950, **72**, 4842.

210. Wender, I., Sternberg, H. W. and Orchin, M., J. Amer. Chem. Soc., 1953, **75**, 3041.

211. Wender, I., Storch, H. H. and Orchin, M., J. Amer. Chem. Soc., 1953, **75**, 3041.

212. Yokokawa, C., Watanabe, Y. and Takegami, Y., Bull. Chem. Soc. Japan, 1964, **37**, 677.

213. Ziesecke, K. H., Brennstoff-Chem., 1952, **33**, 385.

CHAPTER 7

CARBOXYLATION REACTIONS

A large variety of carboxylic acids and their derivatives can be prepared from the reaction of carbon monoxide with suitable substrates in the presence of a transition metal compound. The feature common to all these reactions is the formation of an acyl-metal derivative which subsequently undergoes hydrolytic cleavage, e.g.

$$R.CO.Co(CO)_4 + H_2O \longrightarrow R.CO.OH + H\ Co(CO)_4$$

Although the formation of an acyl-metal species has been detected in the carboxylation of allyl halides[86] no detailed study has been made of the subsequent hydrolysis.

In view of the preceding discussion on hydroformylation the most obvious substrate for hydrocarboxylation is an olefin, Table 7.1. The conditions generally used for this purpose entail the reaction of the olefin with carbon monoxide under aqueous acidic conditions in the presence of a nickel salt. Temperatures in the region of 250°C and pressures of 200 atmospheres are typical. The nickel salt, preferably the iodide, is apparently converted into nickel carbonyl in situ. If a stoichiometric amount of nickel carbonyl is used less stringent conditions are necessary and temperatures around 160°C at pressures of about 50 atmospheres of carbon monoxide are normally employed. However, strained olefins, particularly bicyclo[2, 2, 1]heptene derivatives, undergo hydrocarboxylation at atmospheric pressure and temperatures around 50°C[14]. It has recently been reported that the hydrocarboxylation of olefins is catalysed by ultraviolet light, which permits the reaction to be carried out under ambient conditions[77].

The stereochemistry of hydrocarboxylation has only been examined in one case, namely that of bicyclo[2, 2, 1]heptene (1) which is converted by nickel carbonyl in a deuterated solvent into 3-*exo*-deuterobicyclo[2, 2, 1]heptane-2-*exo*-carboxylic acid (2)[14]. This implies a

TABLE 7.1 Hydrocarboxylation of Monoolefins

Olefin	Catalyst	Products*	Yield, %	References
Ethylene	Co	Propionic acid, diethylketone	85	81, 132, 133
	Ni	Propionic acid	89	81, 82, 83, 105, 118, 133
	Ru	Propionic acid	—	4, 95
	Rh	Propionic acid	—	4, 168
	Pd	Propionic acid, γ-ketocaproic acid	—	168
	Os	Propionic acid	—	95
	Pt	Propionic acid	—	95
Propylene	Co	n-Butyric and iso-butyric acids (3-4:1)	92	81, 103, 111, 119, 120, 121
	Ni	n-Butyric and iso-butyric acids (0.75:1)	—	83, 120, 133
	Pd	n-Butyric and iso-butyric acids (1:2)	—	168
	Pt	Butyric acids	—	95
But-1-ene	Ni	n-Valeric and iso-Valeric acids	—	133
cis-But-2-ene	Pt	2-Methylbutyric acid	—	95
iso-Butene	Ni	Isovaleric acid and trimethylacetic acid (6:1)	—	133
Pent-1-ene	Co	C_6 carboxylic acids	—	103
Hex-1-ene	Co	Heptanoic and 2-methylhexanoic acids	70	66
Oct-1-ene	Ni	Nonanoic and 2-methyloctanoic acids (1:1.7)	70	77, 133, 138
cis-Oct-4-ene	Ni	2-Propylhexanoic acid	17	77

150

Substrate	Catalyst	Product	Yield	Ref.
2-Ethylhex-1-ene	Ni	2-Ethylheptanoic acid	60	133
Dodec-1-ene	Ni	Tridecanoic and 2-methyldodecanoic acids	28	133
Octadec-1-ene	Ni	2-Methyloctadecanoic acid	67	133
Cyclopentene	Ni	Cyclopentanecarboxylic acid	22	14
Cyclohexene	Co	Cyclohexanecarboxylic acid	86	66, 70, 71, 103
	Ni	Cyclohexanecarboxylic acid	83	132, 133, 77
	Pd	Cyclohexanecarboxylic acid	—	168
	Pt	Cyclohexanecarboxylic acid	—	95
Cyclooctene	Ni	Cyclooctanecarboxylic acid	31	136
Cyclododecene	Co	Cyclododecanecarboxylic acid	60	79
	Pd	Cyclododecanecarboxylic acid	93	172
Bicyclo-[2, 2, 1]heptene	Ni	Bicyclo[2, 2, 1]heptane-2-carboxylic acid	80	14
Styrene	Pd	α-and β-phenylpropionic acids (2:1)	—	168
Vinyl chloride	Pd	Propionic and traces of α-chloropropionic acids	—	168
Dihydrofuran	Ni	Tetrahydrofurancarboxylic acid	—	133
Methyl vinyl ketone	Ni	Levulinic and 2-methylacetoacetic acids	—	133
Methyl acrylate	Rh	Dimethyl fumarate	—	182
Allyl alcohol	Co or Rh	γ-Butyrolactone (1-2%), propionaldehyde	—	75
Methallyl alcohol	Co or Rh	β-Methyl-γ-butyrolactone	1-2	75
Crotyl alcohol	Co	α-Methyl-γ-butyrolactone (2%), δ-valero-lactone (0.5%)	—	75
Hex-1-en-4-ol	Co	α-Methyl-γ-ethyl-γ-butyrolactone (2%), hexan-3-one (73%)	—	75

(Continued overleaf)

TABLE 7.1 (continued)

Olefin	Catalyst	Products*	Yield, %	References
2-Methylpent-4-en-2-ol	Co	α,γ,γ-Trimethyl-γ-butyrolactone (10%), δ,δ-dimethyl-valerolactone (2%), 2-methylpentenes (44%)	—	75
2,2-Dimethylbut-3-en-1-ol	Co	α,β,β-Trimethyl-γ-butyrolactone (51%), γ,γ-dimethyl-δ-valerolactone (14%)	—	75
2,2,3-Trimethylbut-3-en-1-ol	Co	α,α,β,β-Tetramethyl-γ-butyrolactone (3%). β,γ,γ-trimethyl-δ-valerolactone (25%)	—	75
2-Ethyl-2-methylbut-3-en-1-ol	Co	αβ-Dimethyl-β-ethyl-γ-butyrolactone (40%), γ-ethyl-γ-methyl-δ-valerolactone (13%)	—	75
Hept-1-en-4-ol	Co	α-Methyl-γ-propyl-γ-butyrolactone (2%), heptan-4-one (70%)	—	75
3-Methylhex-5-en-3-ol	Co	α,γ-Dimethyl-γ-ethyl-γ-butyrolactone (29%), δ-ethyl-δ-methyl-δ-valerolactone (6%)	—	75
4,6-Dimethylhept-1-en-4-ol	Co	α,γ-Dimethyl-γ-isobutyl-γ-butyrolactone (10%), δ-isobutyl-δ-methyl-δ-valerolactone (2%)	—	75
Cyclohex-3-enylcarbinol	Co	2-Hydroxymethylcyclohexanecarboxylic acid lactone (16%)	—	75
Allylamine	Co	Pyrrolidone (40-50%), 3,5-dimethyl-2-ethylpyridine (19-9%), 3,5-dimethyl-4-ethylpyridine (9-1%)	—	74
N-Methylallylamine	Co	N-Methylpyrrolidone	78	74
N-Ethylallylamine	Co	N-Ethylpyrrolidone	61	74
N-Isobutylallylamine	Co	N-Isobutylpyrrolidone	61	74

N-Octylallylamine	Co	N-Octylpyrrolidone	84	74
N-Dodecylallylamine	Co	N-Dodecylpyrrolidone	56	74
But-2-enylamine	Co	3-Methylpyrrolidone (36%), piperidinone (4%)	—	74
N-Methylmethallylamine	Co	1,4-Dimethylpyrrolidone	41	74
Cyclohex-3-enylmethylamine	Co	2-Aminomethylcyclohexane carboxylic acid lactam (39%), 3-aminomethylcyclohexane carboxylic acid lactam (31%)	—	74
Diallylamine	Co	N-Allylpyrrolidone at 150°	—	74
		N-Propenylpyrrolidone at 190°	—	74
Acrylamide	Co	Succinimide	82	73
N-Methylacrylamide	Co	N-Methylsuccinimide	65	73
N-Allylacrylamide	Co	N-Allylsuccinimide	57	74
N-n-Butylacrylamide	Co	N-n-Butylsuccinimide	72	73
N-Isobutylacrylamide	Co	N-Isobutylsuccinimide	80	73
N-Hexylacrylamide	Co	N-Hexylsuccinimide	77	73
N-Laurylacrylamide	Co	N-Laurylsuccinimide	55	73
N-Phenylacrylamide	Co	N-Phenylsuccinimide	64	73
N-p-Chlorophenylacrylamide	Co	N-p-Chlorophenylsuccinimide	65	73
N-2,6-Dichlorophenylacrylamide	Co	N-2,6-Dichlorophenylsuccinimide	44	73
N-Benzylacrylamide	Co	N-Benzylsuccinimide	92	73
N-Carbethoxymethylacrylamide	Co	N-Carbethoxymethylsuccinimide	70	73
Crotonamide	Co	α-Methylsuccinimide (68%), β-methylglutari-mide (19%)	—	73

(Continued overleaf)

TABLE 7.1 (continued)

Transition Metal Intermediates

Olefin	Catalyst	Products*	Yield, %	References
Methacrylamide	Co	α-Methylsuccinimide	68	73
N-Methylmethacrylamide	Co	N-Methyl-α-methylsuccinimide	70	73
N-n-Butylmethacrylamide	Co	N-n-Butyl-α-methylsuccinimide	74	73
N-Benzylmethacrylamide	Co	N-Benzyl-α-methylsuccinimide	76	73
β-Methylcrotonamide	Co	β-Methylglutarimide	67	73
α,α-Dimethylvinylacetamide	Co	α,α-Dimethylglutarimide	58	73
α,α-Dimethylpropenylacetamide	Co	α,α,α'-Trimethylglutarimide	64	73
Cyclohex-1-enecarboxamide	Co	Hexahydrophthalimide	91	73
Cyclohex-1-enylacetamide	Co	1,3-Dioxohexahydroisoquinoline	41	73
Cinnamide	Co	α-Phenylsuccinimide	32	73

*In many cases the carboxylic acids have been obtained as esters or other derivatives.

cis-addition to the *exo*-side of the double-bond. The absence of deuterium incorporation into any other part of the molecule provides incidental support for a non-carbonium ion hydrocarboxylation mechanism.

Ketone formation is sometimes observed during hydrocarboxylation reactions. Ethylene has been found to yield appreciable amounts of diethyl ketone in addition to propionic acid[81, 133]. The use of complex nickel cyanides as catalysts appears to accentuate ketone formation, as under these conditions ethylene also produces octa-3, 6-dione, undeca-3, 6, 9-trione, tetradeca-3, 6, 9, 12-tetraone, 4-keto-hexanoic acid and 3, 6-diketooctanoic acid[135a]. The hydrocarboxylation of bicyclo[2, 2, 1]heptadiene with nickel carbonyl in dioxan-water-acetic acid produces increasing amounts of the ketone (3) as the proportion of water is decreased[14]. This is in accord with competition between water and olefin for reaction with an acyl-nickel intermediate.

(3)

The use of nickel catalysts results in the formation of large amounts of branched chain acids from terminal olefins. For example propylene gives a mixture of *n*- and *iso*-butyric acids in a ratio of about 0.75.[120] In contrast dicobalt octacarbonyl which is an equally effective catalyst gives the same acids in a ratio of about 3.0. The ratio increases with carbon monoxide pressure[119].

A kinetic study[71] of the hydrocarboxylation of cyclohexene using dicobalt octacarbonyl shows the rate to be proportional to the concentration of water and the half power of the olefin concentration. The maximum rate at 156°C is attained at carbon monoxide pressures between 165 and 210 atmospheres. At lower pressures the rate is directly proportional to the carbon monoxide pressure and at higher pressures decreases proportionately becoming inversely proportional about 340 atmospheres. The rate also depends on the cobalt concentration but to a less than proportional extent. The rate of hydrocarboxylation reactions is greatly increased by the introduction of a small amount of hydrogen[103, 111], and the initial rate becomes first order in olefin[111]. The reaction is catalysed by amines in the decreasing order pyridine > quinoline > α-picoline > dimethylaniline[111]. Triethylamine, piperidine and benzylamine practically inhibit the reaction.

Of other metal catalysts that have been investigated iron compounds are much less effective than those of nickel or cobalt. Palladous chloride catalyses the hydrocarboxylation reaction giving a very

high proportion (2:1) of *iso*- to *n*-butyric from propylene[168].
Rhenium[168], rhodium[4, 168, 182], and ruthenium[4, 95] are also credited
with catalytic activity. The very active catalyst systems obtained
from the combination of stannous chloride with a platinum, ruthenium
or osmium compound also catalyse hydrocarboxylation of olefins[95].

By conducting these hydrocarboxylation reactions in the presence of
an appropriate alcohol, thiol, amine, carboxylic acid or hydrogen
chloride the corresponding ester, thioester, amide, anhydride or acid
chloride is obtained[131, 133]. An interesting extension is to generate
the amine also from the olefin in situ[151]. For example, cyclohexene
is converted into N-(cyclohexylmethyl)cyclohexanecarboxamide by
reaction with carbon monoxide, hydrogen and ammonia in the presence
of cobalt. The conversion can be pictured as occurring through initial hydroformylation, imine formation and reduction to cyclohexyl-
methylamine, which then reacts with cyclohexanecarbonyl cobalt
tetracarbonyl.

$$C_6H_{10} + CO + H_2 \xrightarrow{Co} C_6H_{11} CHO \xrightarrow{NH_3} C_6H_{11}CH = NH \xrightarrow{H_2}$$

$$C_6H_{11} CH_2NH_2 \xrightarrow{C_6H_{11} CO.Co(CO)_4} C_6H_{11}CO.NH.CH_2C_6H_{11}.$$

In the presence of iron, such as the autoclave, much tri-(cyclohexylmethyl)amine is formed, presumably by successive condensation of
cyclohexylaldehyde with mono- and di-(cyclohexylmethyl)amine and
reduction of the resulting Schiff's bases. Ethylene in addition to N-
n-propylpropionamide and propionamide yields 3,5-dimethyl-2-
ethylpyridine (5) and 3,5-dimethyl-4-ethylpyridine (6). The formation of these latter products probably occurs by the route outlined
on page 157.

(4)   (5)   (6)

The intramolecular equivalent of amide and ester formation has been
effected by hydrocarboxylation of unsaturated amines and alcohols.
Allylamine is converted into pyrollidone (4) and smaller amounts of
3,5-dimethyl-2-ethylpyridine (5) and 3,5-dimethyl-4-ethylpyridine
(6) using dicobalt octacarbonyl[74, 76]. Only traces of these are formed
using nickel carbonyl, most of the allylamine being converted into
N-allylformamide. The proportions of the pyridine (5) are even

greater using rhodium carbonyl or iron pentacarbonyl[76]. In the latter case the formation of (5) has been shown to follow the path

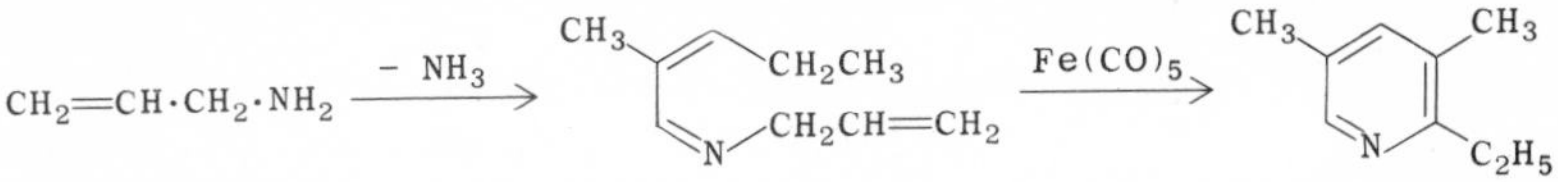

$$CH_2=CH\cdot CH_2\cdot NH_2 \xrightarrow{-NH_3}$$

$$\xrightarrow{Fe(CO)_5}$$

In suitable cases where double bond migration can occur piperidones are formed in addition to pyrrolidones[74].

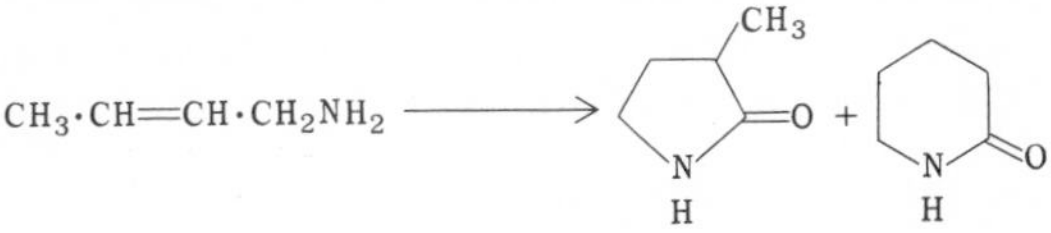

$$CH_3\cdot CH=CH\cdot CH_2NH_2 \longrightarrow$$

36%   4%

Bicyclic lactams have also been obtained; the apparent double bond migration in the following case is noteworthy.

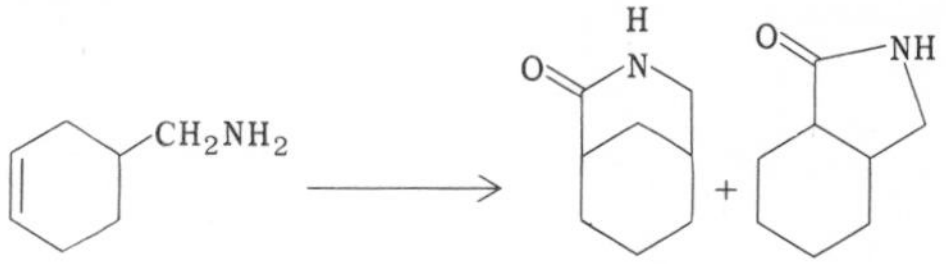

31%   39%

In an analogous reaction unsaturated amides give rise to cyclic imides[72,73]. For example acrylamide is converted into succinimide in 82% yield using dicobalt octacarbonyl.

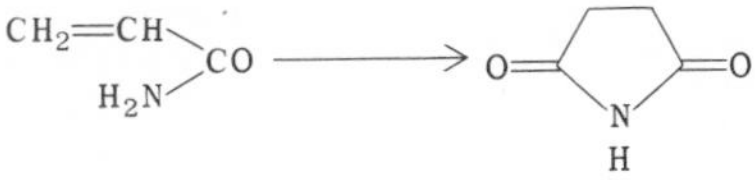

When an alkyl group is present on the β-position of the acrylamide the equilibrium

$$CH_3\cdot CR = CH\ CONH_2 \rightleftharpoons CH_2 = CR\cdot CH_2\cdot CONH_2$$

comes into operation and both succinimide and glutarimide derivatives are formed, the actual proportions depending upon R.

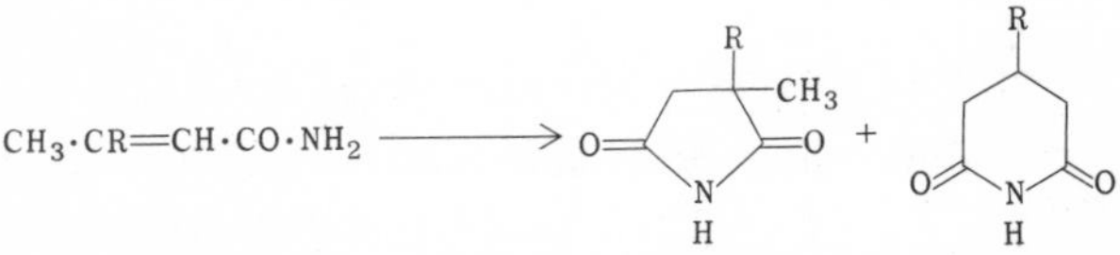

$$CH_3\cdot CR=CH\cdot CO\cdot NH_2 \longrightarrow$$

	68%	19%
R = H		
R = CH$_3$	0	67%

A particularly illuminating case is supplied by N-allyl-acrylamide which produces only N-allylsuccinimide and none of the alternative N-acrylylpyrrolid-2-one[74].

Carboxylation of unsaturated alcohols to give γ- or δ-lactones proceeds in only low yield as the result of side-reactions[75]. The most important of these with allyl alcohol is isomerisation to propionaldehyde. Allyl alcohol gives yields of 1-2% of γ-butyrolactone using dicobalt octacarbonyl or rhodium trichloride as the catalyst, but none is obtained with nickel iodide, iridium trichloride or iron pentacarbonyl. As with allylamines double-bond migration in appropriate allyl alcohols leads to formation of both γ- and δ-lactones.

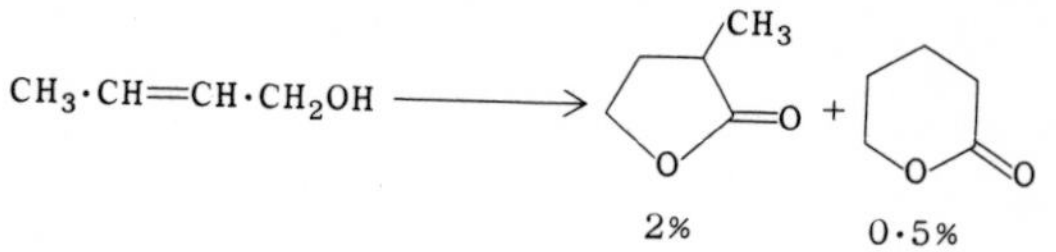
CH3·CH=CH·CH2OH
CH3
O
O
2%
0·5%

However the unsaturated alcohol (**7**) gives only the γ-lactone (**8**).

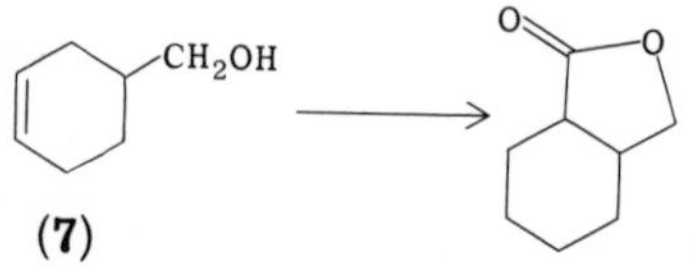
CH2OH
O
O

(**7**) (**8**)

Better yields of lactones are obtained from but-3-enyl alcohols where isomerisation to a saturated carbonyl compound is impossible, e.g.

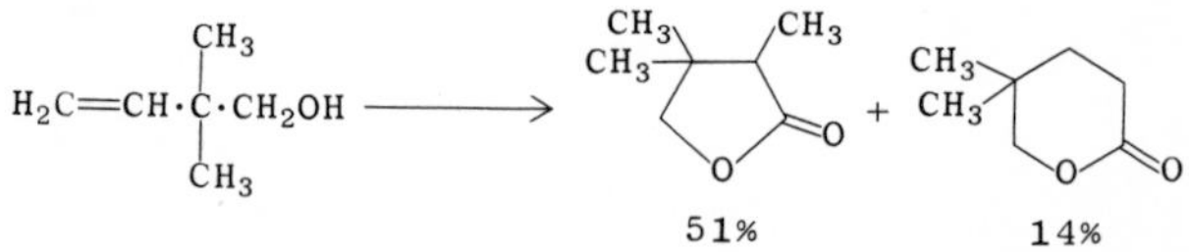
CH3
H2C=CH·C·CH2OH
CH3
CH3
CH3
CH3
CH3
CH3
51%
14%

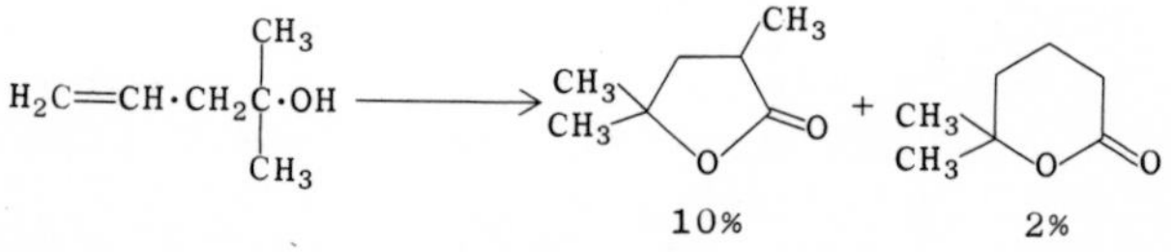
CH3
H2C=CH·CH2C·OH
CH3
CH3
CH3
CH3
CH3
CH3
10%
2%

In the latter case extensive hydrogenolysis occurs.

The normal olefin carboxylation procedure can also be applied to saturated alcohols but a much higher reaction temperature is necessary. Secondary and tertiary alcohols react at 275°C and primary at 300°C[1]. Large proportions of branched chain acids are obtained from primary alcohols, Table 7. 2, and it seems that the reaction may

TABLE 7.2 Carboxylation of Alcohols, Ethers and Lactones

Substrate	Catalyst	Products	Yield, %	References
Methanol	Co	Acetic acid	20	22, 78, 89, 112, 113, 114
Ethanol	Ni	Acetic acid	87	83, 134
n-Propanol	Ni	Propionic acid	69	5, 6, 134
iso-Propanol	Ni	n-Butyric and isobutyric acid (4:1)	70	134
n-Butanol	Ni	iso-Butyric acid	—	83
Butan-2-ol	Ni	α-Methylbutyric acid	47	1
	Ni	α-Methylbutyric acid	30	1
tert.-Butanol	Ni	Isovaleric and trimethylacetic acid (9:1)	57	134
n-Pentanol	Ni	α-Methylvaleric acid	16	1
neo-Pentanol	Ni	C₆ acids	21	1
2-Methylbutanol	Ni	Dimethylethylacetic acid	35	1
Hexanol	Ni	2-Methylhexanoic acid	55	1
2-Ethylbutan-1-ol	Ni	Diethylmethylacetic acid	40	1
Heptanol	Ni	2-Methylheptanoic acid	33	1
Heptan-2-ol	Ni	2-Methylheptanoic acid	70	1
Octanol	Ni	2-Methyloctanoic acid	30	1
Octan-2-ol	Ni	2-Methyloctanoic acid	76	1
3-Cyclohexylpropanol	Ni	3-Cyclohexyl-2-methylpropionic acid	36	1

(Continued overleaf)

TABLE 7.2 (continued)

Substrate	Catalyst	Products	Yield, %	References
Cyclopentanol	Ni	Cyclopentanecarboxylic acid	84	1
4-Methylcyclohexanol	Ni	4-Methylcyclohexanecarboxylic acid	53	1
Decal-2-ol	Ni	Decahydro-1 and 2-naphthoic acids	77	1
2-Phenylethanol	Ni	Ethylbenzene	12	1
3-Phenylpropanol	Ni	n-Propylbenzene	40	1
4-Phenylbutanol	Ni	n-Butylbenzene	40	1
Ethane-1,2-diol	Ni	Succinic acid	—	134
Butane-1,3-diol	Ni	Adipic and α-methylglutaric acids (2:3)	12.5	134
Butane-1,4-diol	Ni	Adipic acid	70	58, 134
Pentane-1,5-diol	Ni	Pimelic acid	—	134
	Ni	α-Methylvaleric acid	10	1
Hexane-1,6-diol	Ni	Suberic acid	—	134
	Ni	2-Methylhexanoic acid	30	1
Decane-1,10-diol	Ni	Decane-1,10-dicarboxylic acid	54	134
Dodecane-1,12-diol	Ni	Dodecane-1,12-dicarboxylic acid	57	134
Tetradecane-1,14-diol	Ni	Tetradecane-1,14-dicarboxylic acid	62	134
Dimethyl ether	Ni	Acetic acid	—	134
Diethyl ether	Ni	Propionic acid	82	5, 83
Propylene oxide	Co	β-Hydroxybutyric acid	40	69
Tetrahydrofuran	Co	δ-Valerolactone, C₅ and C₆ diols, 2-hydroxy-methyl-tetrahydropyran	—	135

	Catalyst	Products	Yield (%)	Reference
	Ni	Adipic acid (74%), valeric acid, δ-valerolactone, α-methylglutaric acid, α-ethylsuccinic acid, α- and γ-methylbutyrolactones	—	5, 135
2-Methyltetrahydrofuran	Ni, Pt	α-Methylbutyric (46%) and adipic (3%) acids	—	80
2,5-Dimethyltetrahydrofuran	Ni	α-Methyladipic acid	—	135
Thiophane	Ni	α,α'-Dimethyladipic acid	—	135
Tetrahydropyran	Ni, Pt	Adipic acid	28	90
Dioxan	Ni	Pimelic acid	—	135
Butyrolactone	Ni	Succinic acid	—	135
α-Methylbutyrolactone	Ni	Glutaric acid	—	135
δ-Valerolactone	Ni	α-Methylglutaric acid	—	135
	Ni	Adipic and valeric acids	—	135

proceed via formation of the olefin. Phenyl substituted alcohols appear to undergo preferential hydrogenolysis to the parent hydrocarbon; for example 4-phenylbutanol gives phenylbutane. It is interesting that diols appear to give mainly straight chain acids.

The problem of isomer formation encountered with the foregoing hydrocarboxylation reactions can be avoided by using the catalytic reaction of an alkyl halide, sulphate or sulphonate with tetracarbonylcobaltate anion in the presence of carbon monoxide and a base[87, 88], Table 7. 3. Reaction temperatures below 50°C can be used with reactive alkylating agents and avoids isomerisation which occurs at higher temperatures. The reaction involves the formation of acylcobalt tetracarbonyls by way of carbonylation of the intermediary alkylcobalt tetracarbonyls. A strong base such as dicyclohexylethylamine catalyses the alcoholysis of the acylcobalt tetracarbonyl, while forming a stable salt with cobalt hydrotetracarbonyl. The latter is thus prevented from reducing the acylcobalt tetracarbonyls. An intramolecular version has produced an unsaturated lactone, in addition to a π-allylic cobalt complex[85]. Introduction of a primary amine into the carboxylation mixture yields the appropriate amide[87, 88]. Disodium iron tetracarbonylate is also able to effect similar carboxylations.

Simple ethers have rarely been subjected to hydrocarboxylation reactions[6,134], but cyclic ethers have been more widely investigated. Five and six-membered ethers are converted into dicarboxylic acids by way of lactones, Table 7. 2, some hydrogenolysis also occurs giving rise to saturated monocarboxylic acids. Trimethylene oxide reacts with cobalt hydrotetracarbonyl yielding firstly γ-hydroxybutyrylcobalt tetracarbonyl (9), which is converted into γ-butyrolactone by treatment with the strong base dicyclohexylethylamine[85].

TABLE 7.3 Carboxymethylation of Alkyl Halides[87, 88]

Alkyl halide	Products	Yield, %
Methyl iodide	Methyl acetate	80
Methyl p-toluenesulphonate	Methyl acetate	33
Amyl iodide	Methyl hexanoate	33
1-Chlorooctane	Methyl nonanoate (28%), methyl 2-methyloctanoate (6%)	—
1-Iodooctane	Methyl nonanoate	56
2-Iodooctane	Methyl 2-methyloctanoate	41
Allyl bromide	Methyl butenoate	33
Benzyl bromide	Methyl phenylacetate	25
α,α'-Dichloro-p-xylene	Dimethyl p-phenylenediacetate	32
α-Chloromethylnaphthalene	Methyl α-naphthylacetate	71
Methyl chloroacetate	Dimethyl malonate	20
Methyl α-bromopropionate	Dimethyl methylmalonate	18

Epoxides react with cobalt hydrotetracarbonyl in the presence of alcohols forming esters of β-hydroxyacids[155, 156, 157]. The order of reactivity is cyclohexene oxide, styrene oxide > propylene oxide > ethylene oxide $\gg$ epichlorohydrin[156]. The reaction rate is increased by the presence of alcohols and to a lesser extent by ketones, ethers and esters[155]. Minor amounts of copper compounds have also been found to promote the reaction[157]. The same conversion of epoxides into β-hydroxyesters can be effected catalytically using dicobalt octacarbonyl[69, 146]. This reaction also results in the formation of ketones as a result of the isomerisation of the epoxides by dicobalt octacarbonyl[68], cf. p. 77

Esters of β-chlorocarboxylic acids can be obtained by carbonylation of olefin-palladium complexes in benzene solution[167, 169], Table 7. 4. The complexes can be prepared in situ. The reaction can be pictured as proceeding as shown in Figure 7. 1, by analogy with the mechanism of palladium catalysed oxidation of olefins (Chapter 4). The formation of straight chain ester from terminal olefins is noteworthy.

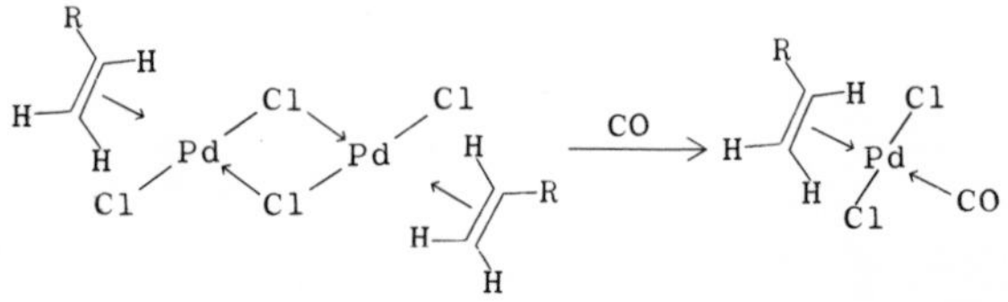

$$\xrightarrow{CO} Cl.CHR.CH_2Pd(CO)_2\ Cl \xrightarrow{CO} Cl.CHR.CH_2CO.Pd(CO)_2Cl$$

$$\longrightarrow Cl.CHR.CH_2COCl + Pd(CO)_x$$

FIGURE 7. 1

As shown the initial product of reaction is the acyl chloride which can be converted into the ester by treatment with an alcohol. If the reaction is carried out in an alcohol saturated with hydrogen chloride the corresponding β-alkoxy ester is obtained[171].

The reaction of cyclopropane with carbon monoxide in benzene in the presence of palladous chloride yields α, β and γ-chlorobutyryl chlorides in a ratio of 5:1:2. [170] The β-chlorobutyryl chloride is formed from propylene, which in turn is formed by isomerisation of cyclopropane under the reaction conditions. The predominance of α-chlorobutyryl chloride is surprising as simple ring opening should give the γ-isomer. Another unexpected product is n-propylbenzene which is presumably formed by a Friedel-Crafts reaction but this only occurs in the presence of carbon monoxide.

TABLE 7.4 β-Substituted Carboxylic Acids from Olefins

Olefin	Products	Yield, %	References
Ethylene	β-Chloropropionyl chloride	37	17
	Ethyl β-chloropropionate	41	167, 169
	Methyl β-methoxypropionate	—	171
	Ethyl β-ethoxypropionate	—	171
	Isopropyl β-iso-propoxypropionate	—	171
Propylene	β-Chlorobutyryl chloride	—	17
	Methyl β-chlorobutyrate	27	167, 169
	Methyl β-methoxybutyrate	—	171
	Ethyl β-ethoxybutyrate	—	171
But-1-ene	Methyl β-chloropentanoate	11	167, 169
Pent-1-ene	Methyl β-chlorohexanoate	10	167, 169
Hex-1-ene	Methyl β-chloroheptanoate and mostly other chloro-esters	—	167, 169
cis-But-2-ene	Methyl α-methyl-β-chlorobutyrate	13	169
trans-But-2-ene	Methyl α-methyl-β-chlorobutyrate	6	169
Isobutene	Methyl β-chloroisovalerate	12	169
Cyclohexene	Methyl 2-chlorocyclohexanecarboxylate	36	169
	Ethyl 2-ethoxycyclohexanecarboxylate	—	171
Vinyl chloride	Methyl β,β-dichloropropionate	5	167, 169
Allyl chloride	Methyl β,γ-dichlorobutyrate	5	169

TABLE 7.5 Hydrocarboxylation of Di- and Tri-olefins

Olefin	Catalyst	Products	Yield, %	References
Butadiene	Ni	2-(3-cyclohexen-1-yl)propionic acid, 2(carboxycyclohexyl)propionic acid.	—	132
	Rh	Pent-4-enoic acid (41%), pent-3-enoic acid (28%) and pent-2-enoic acid (3%).	—	181
	Pd	Pent-3-enoic acid	73	19, 20, 160
	Pt	Pent-3-enoic acid	—	95
Penta-1,3-diene	Pd	2-Methylpent-3-enoic acid	34	19
Penta-1,4-diene	Pd	Expected acids and 2-carboxymethylcyclopentanone (5-10%)	—	20
Isoprene	Pd	3-Methylpent-3-enoic (15%), 4-methylpent-3-enoic (38%) and 4-methylpent-4-enoic (10%) acids	—	19
Hexa-1,5-diene	Ni	2-Methylhex-5-enoic acid	20	132, 133
	Pd	2-Carboxymethyl-5-methyl cyclopentanone	40-50	20
2,3-Dimethylbutadiene	Pd	3,4-Dimethylpent-3-enoic acid	50	19
Cyclopentadiene	Ni	Tricyclo-[5, 2, 1, 0², 6]dec-3- or -4-ene-exo-8-carboxylic acid	69	14
	Pd	Cyclopent-2-ene-1-carboxylic acid	73	19
Cyclohexa-1,3-diene	Ni	Cyclohex-3-ene-1-carboxylic acid	—	14
	Pd	Cyclohex-2-ene-1-carboxylic acid	80	161
Cyclohexa-1,4-diene	Ni	Cyclohex-3-ene-1-carboxylic acid	29	14
Cycloocta-1,3-diene	Pd	Cyclooct-2-ene-1-carboxylic acid (14%), diacids (3%)	—	20, 161

Substrate	Catalyst	Product	Yield	Ref.
Cycloocta-1,5-diene	Pd	Cyclooct-4-enecarboxylic acid and/or diacids	—	20, 159, 161
cis, trans-Cyclododeca-1,5,9-triene	Co	Cyclododecadienecarboxylic acid	13	142
	Co	Cyclododecanecarboxylic acid	88	142
	Pd	Cyclododeca-5,9-dienecarboxylic and/or cyclododec-9-ene-1,5-dicarboxylic acids	—	9, 172
Bicyclo[2,2,1]heptadiene	Ni	Bicyclo[2,2,1]hept-2-ene-5-exo-carboxylic acid, bicyclo[2,2,1]heptane-2,5 or 2,6-dicarboxylic acid, dibicycloheptenyl ketone	—	14
Bicyclo[2,2,2]octadiene	Ni	Bicyclo (2,2,2]oct-5-ene-2-carboxylic acid, diacids and dibicyclooctyl ketone	—	14
Dicyclopentadiene	Ni	Tricyclo[5,2,1,0^{2,6}]dec-3 or 4-ene-exo-8-carboxylic acid	80	14
1,2,3,4,7,7-Hexachloro-bicyclo[2,2,1]heptadiene	Ni	1,4,5,6,7,7-Hexachlorobicyclo[2,2,1]hept-5-ene-2-carboxylic acid	—	14
Isodrin	Ni	Dihydroisodrincarboxylic acid	71	14
Aldrin	Ni	Dihydroaldrincarboxylic acid	73	14

Hydrocarboxylation of non-conjugated diolefins results in the formation of mono-unsaturated acids or diacids, Table 7. 5. Occasionally hydrogenation is observed as in the hydrocarboxylation of cyclodeca-1, 5, 9-triene using a cobalt catalyst. When acetone is used as the solvent cyclododecadiene carboxylic acid is obtained but in methanol, which can apparently act as a hydrogen source, cyclododecanecarboxylic acid is formed[142]. Hydrocarboxylation of *cis, trans, trans*-cyclododeca-1, 5, 9-triene with a palladium catalyst occurs firstly at a *trans* double bond and then randomly at the remaining *trans* or *cis* double bond[172]. If the double bonds are suitably juxtaposed intramolecular reactions are observed. Penta-1, 4-diene gives the keto-acid (**10**)(10%) in addition to the normal esters[20]. Hexa-1, 5-diene produces even higher yields (40-50%) of (**11**) but

hepta-1, 6-diene forms only traces of ketonic products. Cycloocta-1, 5-diene under hydroxylic conditions yields only the expected esters[20], but in tetrahydrofuran the bicyclic ketone (**12**) is formed in 40-45% yield[21].

(**12**)

Conjugated 1, 3-dienes on hydrocarboxylation yield β, γ-unsaturated acids resulting from a formal 1, 4-addition of the elements of formic acid. The most frequently used catalysts are palladium compounds, the most effective[19] is bis(tributylphosphine)palladium iodide. The process appears to proceed by way of π-allylpalladium complexes such as (**13**) formed from butadiene and palladous chloride in benzene. Carbonylation of complex (**13**) gives the products shown[160, 163].

Carbonylation of the complex (**14**), derived from reaction of palladous chloride with isoprene in ethanol, gives a large range of products which depend on the reaction conditions, Figure 7. 2. [160] The dicar-

TABLE 7.6 Carboxylation of Allyl Derivatives

Allyl compound	Catalyst	Products	Yield, %	References
Allyl chloride	Ni	Vinylacetic acid, trace of crotonic acid	60	32, 49, 63
	Pd	Vinylacetic acid, traces of crotonic, isobutyric and β-chlorobutyric acids	96	63, 109, 147 158, 164, 165
Allyl p-toluenesulphonate	Pd	Vinylacetic (24%) and crotonic (26%) acids	—	158, 164
Allyl alcohol	Pd	Vinylacetic acid	42	158, 164, 165
Allyl methyl ether	Pd	Vinylacetic acid	—	147
Allyl ethyl ether	Pd	Vinylacetic acid (57%), crotonic acid (8%)	—	158, 164
Allyl phenyl ether	Pd	Vinylacetic (18%) and crotonic (10%) acids	—	158, 164
Allyl acetate	Pd	Vinylacetic acid	80	147, 158, 164
1-Chlorobut-2-ene	Ni	Pent-3-enoic acid	—	27, 32
	Pd	Pent-3-enoic acid	81	63, 109
3-Chlorobut-1-ene	Pd	Pent-3-enoic acid	84	63, 109, 147
But-1-en-3-ol	Pd	Pent-3-enoic acid	39	164
But-2-en-1-ol	Pd	Pent-3-enoic acid	42	158, 164
But-2-en-1-ol acetate	Pd	Pent-3-enoic acid	64	164
Methallyl chloride	Pd	3-Methylbut-3-enoic acid (78%), 3-methylbut-2-enoic acid (7%)	—	63, 109, 147, 158
Pent-3-en-2-ol acetate	Pd	2-Methylpent-4-enoic acid	39	158, 164
Cyclohex-2-enyl chloride	Pd	Cyclohex-2-enecarboxylic acid	—	147
Cyclohex-2-enyl bromide	Pd	Cyclohex-2-enecarboxylic acid	30	158, 164
Cyclooct-2-enyl chloride	Pd	Cyclooct-2-enecarboxylic acid	—	147
Cyclooct-2-enyl bromide	Pd	Cyclooct-2-enecarboxylic acid	36	161
3-Chloroallyl chloride	Ni	4-Chlorobut-3-enoic acid	5	43
1-Chloro-4-cyanobut-2-ene	Ni	5-Cyanopent-3-enoic acid	63	27, 32, 35

169

boxylic acids undoubtedly have their origin in the ability of $\alpha\beta$-
and $\beta\gamma$-unsaturated carboxylic esters to form π-allylic palladium
complexes such as (15), which can then undergo further carbonylation[162].

(15)

Allyl halides and similar allylic compounds which form π-allylic
palladium complexes are also readily converted to $\beta\gamma$-unsaturated
acids[63, 110, 164], Table 7.6. Other catalysts such as rhodium trichloride and chloroplatinic acid with stannous chloride have been
used but are stated to be less efficient[158, 164]. Allyl halides are
also converted to $\beta\gamma$-unsaturated acids by nickel carbonyl[32]. Here
also the formation of a π-allylicnickel complex is implicated and the
conversion of (16) into vinylacetic acid by way of the but-3-enoyl-
nickel dicarbonyl bromide species has been demonstrated[86].

(16)

The concurrent allylic coupling reactions which always occur with
nickel carbonyl are minimised by working under a pressure of carbon monoxide.

π-Allylic palladium complexes can also be obtained from allene and
have the structures (17) and (18). Complex (17) on treatment with
carbon monoxide in ethanolic solution produces the expected ethyl
3-chloro-but-3-enoate[175]. Under the same conditions complex (18)
forms at least three esters. Direct carbonylation of allene with palladous chloride yields only diethyl itaconate. Hydrocarboxylation of
allene in methanol using an iron or ruthenium carbonyl catalyst produces methyl methacrylate, dimethyl 2-methylene-3, 3-dimethylglu-
tarate and methyl 3-methylene-1-methylcyclobutane carboxylate[11,
102]. When ruthenium trichloride and pyridine in aqueous solution is
used two lactones (19) and (20) are obtained in addition to methacrylic acid. In methanolic solution the ester (21) is formed instead.
In contrast platinous chloride with stannous chloride gave methyl
methacrylate[95]. Carbonylation of a mixture of allene and acetylene
using ruthenium trichloride and pyridine in water forms a small
amount of 3-methylcyclopent-3-enone[102].

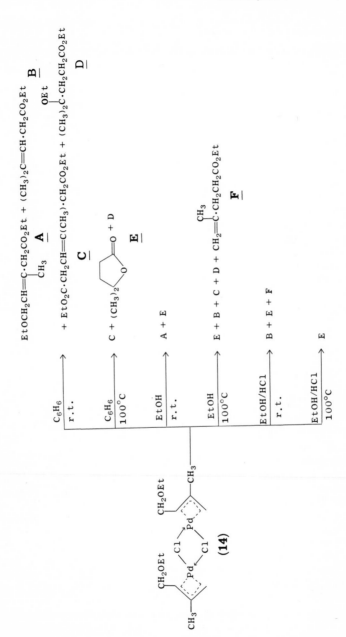

FIGURE 7.2 (Compounds are listed in order of decreasing amount.)

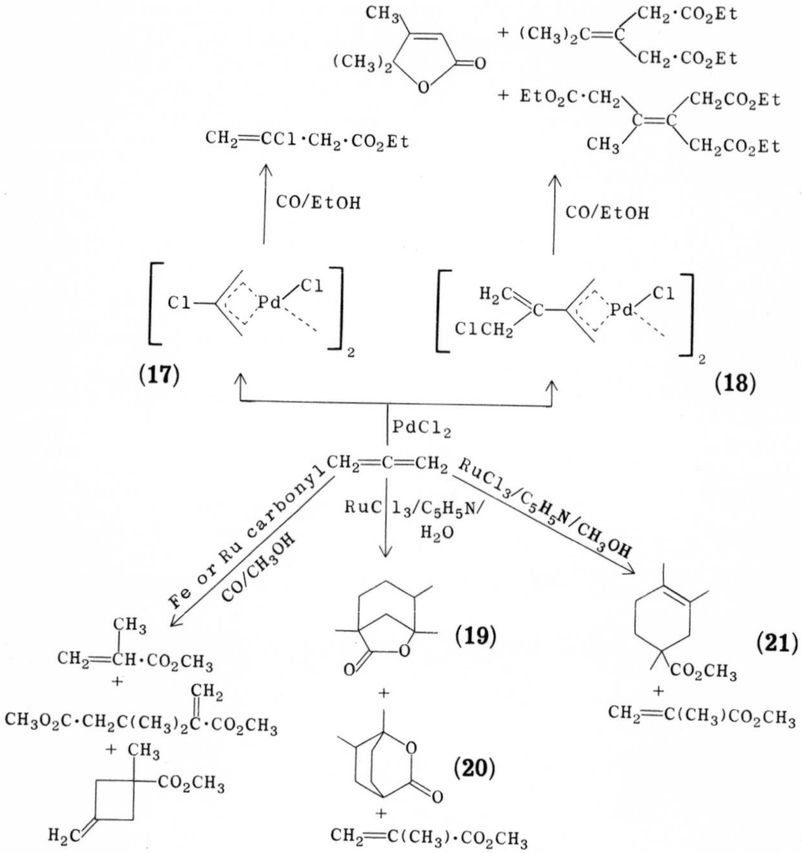
CO/EtOH
CO/EtOH
(17)
(18)
PdCl₂
Fe or Ru carbonyl CO/CH₃OH
RuCl₃/C₅H₅N/H₂O
RuCl₃/C₅H₅N/CH₃OH
(19)
(20)
(21)

FIGURE 7.3

A novel electrolytic method for the conversion of olefins to esters of α, β-unsaturated acids has been discovered[92, 93]. Electrolysis of a methanolic solution of sodium methoxide in the presence of carbon monoxide with platinum cathode and anode generates a solution of an uncharacterised platinum carbonyl species. The platinum cathode is subsequently replaced by a copper one and the electrolysis continued in the presence of the olefin and carbon monoxide. The esters are formed at the anode. In this way, for example, styrene is converted into methyl *trans*-cinnamate (21%), β-methoxy-β-phenylpropionate (13%), β-phenylpropionate (1%) and the dimethyl ether of styrene glycol (23%). Similar products are obtained from α-methyl-styrene and 1, 1-diphenylethylene.

$$C_6H_5CH = CH_2 \longrightarrow$$

$$\longrightarrow C_6H_5.CH = CH.CO_2.CH_3$$

$$\overset{O\ CH_3}{\underset{|}{\longrightarrow C_6H_5.CH.CH_2.CO_2CH_3}}$$

$$\longrightarrow C_6H_5.CH_2.CH_2.CO_2CH_3$$

$$\overset{OCH_3\ OCH_3}{\underset{|\quad\ \ |}{\longrightarrow C_6H_5.CH-CH.CO_2CH_3}}$$

The most extensively investigated source of $\alpha\beta$-unsaturated acids is the hydrocarboxylation of acetylenes, Table 7.7. The fundamental reaction is typified by the exothermic conversion of acetylene to acrylic acid when reacting with nickel carbonyl under aqueous acidic conditions, at about 40°C. The reaction may be carried out with either the stoichiometric quantity of nickel carbonyl or catalytically by formation of the nickel carbonyl in situ from carbon monoxide and a nickel salt. Typical reaction conditions for the latter are 150°C and 30 atmospheres.

A wide variety of solvent systems have been employed successfully[99] including primary, tertiary and secondary alcohols, acetone, methyl ethyl ketone, dioxan, tetrahydrofuran, ethyl acetate, pyridine and anisole. However, water must be present and in anhydrous alcoholic solvents only much reduced yields of acids are obtained. Non-hydroxylic solvents such as dioxan favour side-reactions. Hydrocarboxylation of diphenylacetylene with nickel carbonyl in dioxan containing only a little acidic aqueous ethanol produces a small amount of ethyl *trans*-α-phenylcinnamate, the main product being 2, 3, 4, 5-tetraphenylcyclopent-2-en-1-one (22). [116] Although it was suggested originally that this product arose from reduction of tetra-

TABLE 7.7 Hydrocarboxylation of Acetylenes

Acetylene	Products	Yield, %	References
Acetylene	Acrylic acid	95	13, 65, 66, 67, 130, 143, 177, 179
But-1-yne	But-1-en-2-carboxylic acid	45	177
Propyne	α-Methacrylic acid, small amounts of crotonic acid and isobutyric acid.	50	84, 143, 177
Hex-1-yne	Hex-1-en-2-carboxylic acid	35	96, 177
Oct-1-yne	Oct-1-en-2-carboxylic acid	20	46, 130
Non-2-yne	α-Hexyl-β-methylacrylic and α-methyl-β-hexyl acrylic acid	32.5	130
Dec-5-yne	Dec-5-ene-5-carboxylic acid	52	97
Phenylacetylene	Atropic acid and traces cinnamic acid	48	96, 130
1-Phenylprop-1-yne	α-Methylcinnamic and α-phenylcrotonic acid	54	130
Diphenylacetylene	α-Phenyl*trans*-cinnamic acid	48	97, 116, 130
Prop-2-yn-1-ol	α-Hydroxymethylacrylic acid (58%), *trans*-γ-hydroxycrotonic acid (11%).	—	140
Prop-2-yn-1-ol acetate	3-Acetoxyprop-1-ene-2-carboxylic acid	40	96
But-1-yn-3-ol	3-Hydroxybut-1-ene-2-carboxylic acid	—	130
But-1-yn-3-ol acetate	3-Acetoxybut-1-ene-2-carboxylic acid (35%), 2-methylbut-2-enoic acid (5%)	—	62
But-3-yn-1-ol	4-Hydroxybut-1-ene-2-carboxylic acid lactone	23	96
But-3-yn-1-ol acetate	4-Acetoxybut-1-ene-2-carboxylic acid	47	96
But-3-yn-1-ol tetrahydropyranyl ether	4-Hydroxybut-1-ene-2-carboxylic acid tetrahydropyranyl ether	20	96

But-2-yn-1,4-diol diacetate	1,4-Diacetoxybut-2-ene-2-carboxylic acid	58	97
Pent-1-yn-5-ol p-toluenesulfonate	5-Hydroxy-pent-1-ene-2-carboxylic acid p-toluenesulfonate	1.5	98
Pent-4-yn-2-ol	4-Hydroxypent-1-ene-2-carboxylic acid lactone	30	96
Pent-3-yn-2-ol	4-Hydroxypent-2-ene-2-carboxylic acid	60	97
Hex-1-yn-3-ol acetate	3-Acetoxyhex-1-ene-2-carboxylic acid	48	96
2-Methylpent-4-yn-2-ol	4-Hydroxy-4-methylpent-1-ene-2-carboxylic acid lactone	7.4	96
1-Ethynylcyclohexyl acetate	α-(1-Acetoxycyclohexyl) acrylic acid	3.5	96
1-Phenylprop-2-yn-1-ol acetate	3-Acetoxy-3-phenylprop-1-ene-2-carboxylic acid	50	96
1-Phenylprop-2-yn-1-ol	3-Hydroxy-3-phenylprop-1-ene-2-carboxylic acid	10	96
5-Bromopent-1-yne	5-Bromopent-1-ene-2-carboxylic acid	40	98
Pent-3-yn-2-one	cis and trans-4-ketopent-2-ene-2-carboxylic acid	30	97
Oct-3-yn-2-one	2-Keto-oct-3-ene-4-carboxylic acid	40	97
Pent-4-yn-1-ol	5-Hydroxypent-1-ene-2-carboxylic acid lactone	21	96
Ethyl propiolate	No reaction	—	100
Sodium propiolate	Trans,trans-muconic acid	1	100
Ethyl but-3-ynoate	Ethyl 3-carboxybut-3-enoate	28	100
Ethyl pent-4-ynoate	Ethyl 4-carboxypent-4-enoate	46	100
Ethyl hex-5-ynoate	Ethyl 5-carboxyhex-5-enoate	40	100
Ethyl hept-6-ynoate	α-n-Butyl-β-carbethoxyacrylic acid	37	97, 100
Hept-6-ynoic acid	n-Butylfumaric acid	28	97
Hex-5-ynonitrile	1-Cyanopent-4-ene-4-carboxylic acid	39	100
Diethyl acetylene dicarboxylate	Ethyl mellitate	18	97

phenylcyclopentadienone a more likely mode of formation is the Nazarov type cyclisation of the intermediary divinyl ketone **(23)**.

Most of the hydrocarboxylation reactions of acetylenes reported in the literature have been carried out in aqueous ethanol when the product is the ethyl ester. Ester formation may be minimised by buffering the reaction medium. A wide range of esters may be obtained by conducting the reaction in the presence of the appropriate alcohol or thiol, and amides in the presence of amines[130]. The prime purpose of the acid does not seem to be the production of hydrogen ion concentration since both hydrochloric and acetic acids function equally effectively although trichloroacetic acid, of intermediate dissociation constant, is not[99].

The reaction is characterised by an induction period the length of which depends on the individual acetylene and the reaction temperature. Thus it is one minute for acetylene at 40°C and for diphenylacetylene at 75°C. Alternatively the reaction is catalysed by ultraviolet light[77]. A transient deep brown colouration is frequently observed during the reaction. The efficiency of, and the precise conditions for, the hydrocarboxylation reaction depend on the individual acetylene. Conjugation of the triple bond with a carbonyl group or the presence of sterically large groups adjacent to it hinder the reaction.

The addition of the elements of formic acid ($H-CO_2H$) is always cis[97]. It was originally deduced from the products obtained (Table 7. 7) that the overall addition obeyed the Markownikoff rule and further study has shown this to be correct[15]. Hydrocarboxylation of monosubstituted diphenylacetylenes where the substituents are in the *meta* or *para* position yield both possible isomers (Table 7. 8) whose relative proportions can be quantitatively correlated with the σ-values of the substituents.

However, when the substituent occupies an *ortho* position the carboxylic acid **(25)** predominates irrespective of the electronic properties of X. Hence, it seems that steric effects are more important than electronic ones in determining the direction of hydrocarboxylation of the acetylenic bond, the carboxyl group becoming attached predominantly to the more sterically crowded end.

TABLE 7.8 Hydrocarboxylation of $C_6H_5.C \equiv C.Ar$

Ar	Ar CH = C(C_6H_5)CO_2H, %	C_6H_6.CH = CAr.CO_2H, %
p-$CH_3O.C_6H_4$	37.3	62.7
p-$CH_3.C_6H_4$	46.5	53.5
m-$CH_3O.C_6H_4$	53.0	47.0
p-$Cl.C_6H_4$	61.8	38.2
m-$Cl.C_6H_4$	66.0	34.0
p-$NO_2C_6H_4$	85.0	15.0
o-$CH_3OC_6H_4$	68.0	32.0
o-$CH_3.C_6H_4$	75.0	25.0
o-$Cl.C_6H_4$	62.5	27.5

The possible intermediary formation of a cyclopropenone was originally suggested to account for the products obtained from hydrocarboxylation of acetylenes[130]. While more recent mechanistic studies in the field obviously exclude this suggestion cyclopropenones are converted into acrylic acids in the essential presence of nickel carbonyl[15, 16]. In the case of diphenyl and dibutylcyclopropenone which are decarbonylated to the corresponding acetylenes by nickel carbonyl under anhydrous conditions, it is difficult to exclude decarbonylation as a first step in their conversion to acrylic acids under aqueous acidic conditions. However, such a course is inconceivable in the case of cycloheptenocyclopropenone (26) which is converted into the acrylic acid in good yield, although all attempts to decarbonylate it and trap the transient cycloheptyne under anhydrous conditions have been unsuccessful[15].

(26)

Hydrocarboxylation of diphenylacetylene has also been achieved under alkaline conditions, although in lower yields, using nickel carbonyl in methanolic sodium hydroxide at room temperature. A

30% yield of *trans*-α-phenylcinnamic acid and 58% of 1, 2, 3, 4-tetra-phenylbuta-1, 3-diene were obtained. The reaction apparently in-volves a $[Ni_3(CO)_8]^{2-}$ species[149].

Vinylacetylenes are not very readily carboxylated but the reaction appears to be catalysed by pyridine[98]. The initially formed diene acids rapidly dimerise so that vinylacetylene gives (27)[130, 177, 178], which is known as mikanecic acid and has been isolated from the alkaline hydrolysis of the alkaloid sarracine[61]. The products from pent-3-en-1-yne[98], 3-methylbut-3-en-1-yne[12] and cyclohex-1-enyl-acetylene[98] are probably correctly formulated as (28), (29) and (30) respectively. Attempts to carboxylate diacetylenes have proved unsuccessful[97].

(27) (28)

(29) (30)

Catalysts other than nickel carbonyl have been used for the hydro-carboxylation of acetylenes, but with less satisfactory results from the point of view of acrylic acid formation. Iron pentacarbonyl gives yields of acrylic acid in the region of 28%, [130] whereas nickel car-bonyl gives over 50%. Carboxylation of acetylene in benzene with palladous chloride yields *trans, trans*-muconyl chloride, fumaryl chloride and maleyl chloride[166]. A small amount of *cis, cis*-muconyl chloride may also be isolated, which is isomerised to the *trans, trans*-isomer under the reaction conditions. Hydrocarboxylation with palladous chloride in ethanol produces ethyl acrylate, propionate, maleate, fumarate and succinate[94]. Under similar conditions di-phenylacetylene gives α, β-diphenyl-γ-crotonolactone (31) and small amounts of ethyl diphenylmaleate[173].

(31)

The use of dicobalt octacarbonyl in methanol at 95°C and a carbon monoxide pressure of 250 atmospheres gives dimethyl succinate as the major product[117, 122, 123, 125], or in the presence of aniline the dianilide[124]. Other products in decreasing amounts are methyl acrylate, dimethyl fumarate, dimethyl α-carbomethoxysuccinate, cyclopent-2-enone, dimethyl γ-oxopimelate, methyl γ-oxopimelate, methyl γ-oxohexanoate, dimethyl *trans, trans*-muconate, dimethyl hex-3-enedioate, and cyclopentanone. Dimethyl succinate is not apparently formed by hydrocarboxylation of methyl acrylate as the reaction is too slow[150]. If hydrogen is also present the ester acetal of γ-oxobutyric acid is a major product[2, 59, 60].

A possible origin of these products is suggested by the following observations. Reaction of acetylene with dicobalt octacarbonyl yields the complex (32), which at 70°C and 200 atmospheres takes up three moles of carbon monoxide and forms the complex (33)[150]. This lactone complex is converted into (34) in acidic aqueous acetone solution at room temperature[3]. Treatment of (34) in benzene with a 1:1 hydrogen-carbon monoxide mixture at 110°C and 260 atmospheres in the presence of dicobalt octacarbonyl produced the complex (35), which has also been isolated from the acetylene hydrocarboxylation reaction. A number of the remaining products can be considered as being formed from divinyl ketone (Figure 7.4). Alternatively dimethyl γ-ketopimelate could be formed from methyl acrylate, and methyl γ-ketohexanoate via formation of methyl 3-ketohex-4-enoate and subsequent reduction.

The course of hydrocarboxylation of halogenoacetylenes with aqueous acidic nickel carbonyl depends upon the position of the halogen relative to the triple bond. Where the halogen is remote from the triple bond the reaction follows the normal course[98]. In compounds where the halogen is directly attached to the acetylenic group as in iodo-but-1-yne the product is the dehalogenated acetylene, namely but-1-yne. The position with regard to α-halogenoacetylenes is more complex the products being allenic acids, maleic anhydrides and ketones. In the case of propargyl chloride small amounts of itaconic acid accompany the major product, butadienoic acid, from which it is also

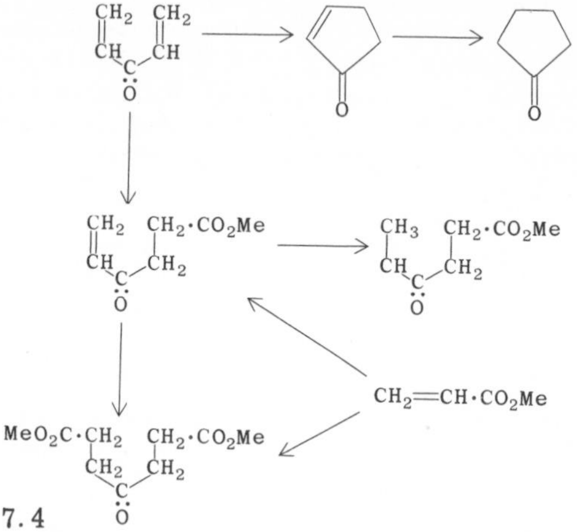

FIGURE 7.4

formed[29, 34]. Carboxylation of propargyl bromide in aqueous hydrochloric acid yields 3-bromobut-3-enoic acid[139]. The formation of some of these products is minimised by the use of α-chloroacetylenes (Table 7.9). The presence of acid is not essential for this reaction and the stoichiometric ratio (nickel carbonyl) : (α-chloroacetylene) : (chloride ion) : (allenic acid) is roughly 2:2:2:1. [101] In the presence of mineral acids such as hydrochloric, the allenic acids lactonise, e.g.

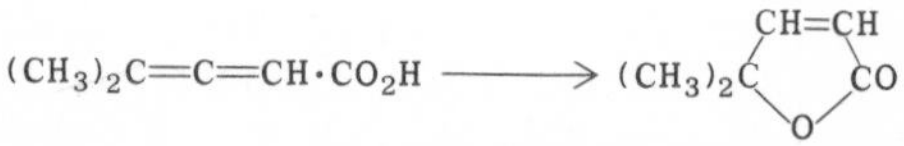

$(CH_3)_2C{=}C{=}CH \cdot CO_2H \longrightarrow$

Comparison of the behaviour of $(CH_3)_2C\,Cl.C{\equiv}CH$ and $(CH_3)_2C{=}C{=}CHCl$ under the same reaction conditions showed that the chloroallene did not react, and thus precluded the reaction proceeding by way of rearrangement of the α-chloroacetylene to the chloroallene[101]. Different acids are obtained from 3-chlorobut-1-yne and 1-chlorobut-2-yne and it is clear that the carboxyl group is introduced at the

$CH_3.CHCl.C{\equiv}CH \longrightarrow CH_3.CH{=}C{=}CH.CO_2H$

$CH_3.C{\equiv}C.CH_2Cl \longrightarrow CH_2{=}C{=}C.CO_2H$ with CH_3 substituent

180

TABLE 7.9 Preparation of Allene Acids

Acetylene	Allene Acid	Yield, %	References
3-Chloroprop-1-yne	Buta-2,3-dienoic acid	6	101, 145
3-Chlorobut-1-yne	Penta-2,3-dienoic acid	10	101
1 Chlorobut-2-yne	2-Methylbuta-2,3-dienoic acid	15	7
3-Chloro-3-methylbut-1-yne	4-Methylpenta-2,3-dienoic acid	45	101
3-Chlorohex-1-yne	Hepta-2,3-dienoic acid	12.5	101
1-Chlorohept-2-yne	2-Butylbuta-2,3-dienoic acid	51	7, 101
1-Bromohept-2-yne	2-Butylbuta-2,3-dienoic acid	11.5	7
1-Iodohept-2-yne	2-Butylbuta-2,3-dienoic acid	22	7
Hept-2-yn-1-ol *p*-toluenesulfonate	2-Butylbuta-2,3-dienoic acid	31	7
2-Chloro-2-methyloct-3-yne	2-Butyl-4-methylpenta-2,3-dienoic acid	13	101
3-Chloro-3-phenylprop-1-yne	4-Phenylbuta-2,3-dienoic acid	12	101

acetylenic carbon atom further from the halogen. This is in accord with the known influence of electronic factors on direction of addition of the elements of formic acid to acetylenes. The loss of chloride ion is probably concerted with the carboxylation reaction, e.g.

$$CH_3.CHCl.C{\equiv}CH \underset{}{\overset{Ni(CO)_4}{\rightleftharpoons}} CH_3.CHCl.C \overset{\diagup}{\underset{\diagdown}{=}} CH$$
$$Ni(CO)_3$$

$$\overset{-Cl^-}{\longrightarrow} CH_3.CH = C = CH.\overset{+}{Ni}(CO)_3 \rightleftharpoons CH_3.CH = C = CH.CO.\overset{+}{Ni}(CO)_3$$

$$\longrightarrow CH_3.CH = C = CH.CO_2H$$

If other α-halogeno- or p-toluenesulphonyloxy-acetylenes are used much lower yields of allenic acids are obtained and two new types of product are observed. These are substituted maleic anhydrides and keto-acids of structure (36).

(36)

The relative amounts depend on the halogen. Thus carboxylation of 1-substituted hept-2-ynes gave the following products[7].

1-Substituent	Allene acid, %	(36), R = n-Bu, %
Cl	40	None
p-$CH_3C_6H_4.SO_3$	31	2
Br	11.5	10
I	20	14

It is thought that small amounts of butylmethylmaleic anhydride may have been missed as all three types of product were observed in the carboxylation of 1-halogenobut-2-yne.

α-Substituent	Allenic acid, %	(36) (R = Me)%	Dimethylmaleic anhydride, %
Cl	15	2	-
Br	-	5	3
I	-	1	14

Maleic anhydride derivatives can also be obtained from acetylenes by reaction with an iron carbonyl in aqueous alkali. Apart from some acrylic acid and hydroquinone derivatives a complex of structure (37) is formed[91, 148]. Oxidation of complex (37, R=R'=CH$_3$) with potassium ferricyanide[57] or nitric acid[23] gives dimethylmaleic anhydride. Complex (37) is converted into complex (38) by oxidation with ferric chloride. This has been accomplished for R = H; R' = H, CH$_3$, C$_2$H$_5$, n-C$_4$H$_9$, C$_6$H$_5$ and for R = R' = H, CH$_3$, C$_2$H$_5$, C$_6$H$_5$.[23] Complex (38, R = R' = C$_6$H$_5$) has been obtained directly from the reaction of diphenylcyclopropenone with diiron nonacarbonyl[15, 16] and complex (38, R = R' = H) from the reaction of acetylene with iron carbonyls[176]. The complex (38, R = R' = H) may be reduced with zinc and acetic acid to the dihydro derivative (39), which can be oxidised to succinic acid[23]. Alternatively (39, R = R' = H) will undergo Diels-Alder addition to cyclopentadiene yielding the adduct (40), which is probably convertable to the diacid.

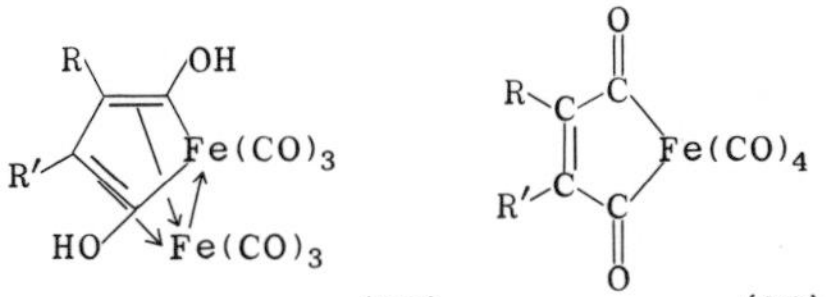

(37)   **(38)**

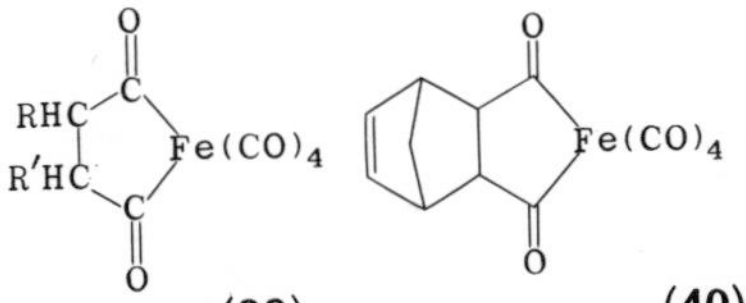

(39)   **(40)**

Good yields of itaconic acids are obtained by platinum catalysed carboxylation of propargyl alcohols[174]. Both propargyl alcohol and chloride give methyl itaconate in about 60% yield. Small amounts of

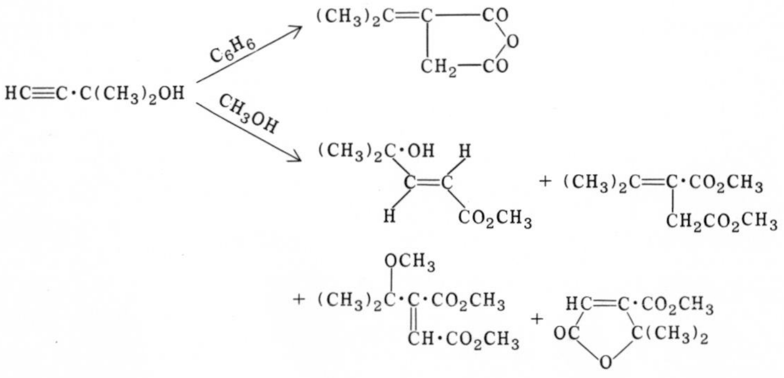

183

TABLE 7.10 Cocarboxylation of Acetylene and Allyl Derivatives

Allyl Compound	Product	Yield, %	References
Allyl chloride	Hexa-2, 5-dienoic acid	41	28, 30, 31, 47, 48, 51, 54
Allyl methyl ether	Hexa-2, 5-dienoic acid	84	51, 53
Allyl butyl ether	Hexa-2, 5-dienoic acid	—	51
Allyl acetate	Hexa-2, 5-dienoic acid	58	51
Allyl acrylate	Hexa-2, 5-dienoic acid (41%), acrylic acid (55%)	—	52
1-Chlorobut-2-ene	Hepta-2, 5-dienoic acid	71	28, 30, 32, 47, 48, 115
3-Chlorobut-1-ene	Hepta-2, 5-dienoic acid	34	28
But-2-enyl methyl ether	Hepta-2, 5-dienoic acid	55	51
But-2-en-1-ol acetate	Hepta-2, 5-dienoic acid	40	51
Methallyl chloride	5-Methylhexa-2, 5-dienoic acid	63	28, 32, 47, 48
Methallyl methyl ether	2-Methylhexa-2, 5-dienoic acid	68	51
4-Chloropent-2-ene	4-Methylhepta-2, 5-dienoic acid	—	28, 32
1-Chloro-3-methylbut-2-ene	6-Methylhepta-2, 5-dienoic acid	36	28, 32
1-Chloro-5, 5-dimethylhex-2-ene	8, 8-Dimethylnona-2, 5-dienoic acid	—	47, 48
1-Chloro-6, 6-dimethylhept-2-ene	9, 9-Dimethyldeca-2, 5-dienoic acid	54	32, 115

1-Chlorocyclopent-2-ene	β-(Cyclopent-2-enyl) acrylic acid	6	28, 32
1-Chlorocyclohex-2-ene	β-(Cyclohex-2-enyl) acrylic acid	—	30, 32
Cinnamyl chloride	6-Phenylhexa-2, 5-dienoic acid	—	28, 30, 32
1, 1-Dichloroprop-2-ene	Hexa-2, 5-dienoic and 6-chlorohexa-2, 5-dienoic acids	—	43
1, 2-Dichloroprop-2-ene	5-Chlorohexa-2, 5-dienoic acid	—	43
cis- or trans-1, 3-Dichloropropene	Hexa-2, 5-dienoic (7%) and 6-chloro-hexa-2, 5-dienoic acids (13%)	—	40, 43
1, 7-Dichlorohepta-2, 5-diene	Trideca-2, 5, 8, 11-tetraenedioic acid	—	115
1, 8-Dichloroocta-2, 6-diene	Tetradeca-2, 5, 9, 12-tetraenedioic acid	19	28, 31, 32
1-Chlorobut-2-en-3-ol acetate	7-Acetoxyhepta-2, 5-dienoic acid	—	115
5-Chloropent-3-enoic acid	Octa-2, 5-dienedioic acid	32	28, 31, 115
1-Chloro-4-cyanobut-2-ene	7-Cyanohepta-2, 5-dienoic acid	65	28, 31, 32, 115
3-Chloro-4-cyanobut-1-ene	7-Cyanohepta-2, 5-dienoic acid	27	28

methyl aconitate are also formed. The proportions of the products from 2-methyl-but-3-yn-2-ol depend on the concentration of hydrogen chloride in the methanol medium. The formation of methyl teraconate and terebate is favoured by higher concentrations. Use of benzene as reaction medium results in the selective formation of teraconic anhydride.

A further development of the hydrocarboxylation of acetylene entails conducting the reaction in the presence of an allyl halide instead of acid[31, 32]. Either nickel carbonyl or Raney nickel and carbon monoxide in the presence of thiourea is used. The product obtained is a cis-2, 5-dienoic acid usually accompanied by small amounts of the 3, 5-isomer as a result of isomerisation (Table 7. 10).

$$CH_2=CH.CH_2Cl + HC\equiv CH \xrightarrow{Ni(CO)_4/H_2O}$$
$$CH_2 = CH.CH_2.CH = CH.CO_2H$$

Experiments using ^{14}C labelled allyl p-toluenesulphonate show that in the course of reaction with acetylene and nickel carbonyl the two ends of the allyl group become equivalent[64]. This observation suggests that a π-allylic nickel complex is involved in the reaction.

Phenol has been obtained as a byproduct of the reaction and it has been found that these dienoic acids can serve as precursors of aromatic rings[36, 37, 38, 39]. For example, hexa-2, 5-dienoic acid is converted into phenyl acetate by heating with acetic anhydride containing zinc chloride, aluminium chloride, stannic chloride or sulphuric acid[36, 38]. A similar cyclisation can be effected by refluxing the acid chloride in xylene in the presence of palladium, iron or silver; possibly as the result of formation of an acyl metal intermediate[39]. Acid catalysed conditions are not satisfactory for ω-alkylhexadienoic acids which are better cyclised using potassium or sodium acetate in acetic anhydride[37].

The amount of water present in the carboxylation mixture has a marked effect on the products obtained from acetylene and allyl chloride. Using acetone, cyclopentanone or acetophenone containing

less than 0.5% water compound (41) is formed, and at very low concentrations the ketone is incorporated giving (42) and (43)[24]. The
similarity to the Reformatsky reaction is stressed by the reaction of
methyl γ-bromocrotonate with nickel carbonyl in acetone yielding
(44) and (45). The reaction of hexa-2,5-dienoyl chloride with acetylene and nickel carbonyl in acetone generates good yields of (41) and
the acid corresponding to (46). [26] There is a marked contrast in
behaviour of allyl halides, such as methyl γ-bromocrotonate, with
electron attracting substituents in their reactions with nickel carbonyl in different solvents. In hydroxylic solvents hydrogenolysis of
the halogen predominates[43] whereas in ketonic solvents methyl γ-
bromocrotonate is carboxylated in the normal way, and with acetylene yields hepta-2,5-dienoic acid and the keto-ester (46). [25]

(41)

(42)   **(43)**

$$CH_2=CH-\underset{\underset{C(CH_3)_2OH}{|}}{CH}-CO_2CH_3 \qquad (CH_3)_2\underset{\underset{OH}{|}}{C}\cdot CH_2\cdot CH=CH\cdot CO_2CH_3$$

(44)   **(45)**

(46)

The formation of cyclic keto-acid and keto-lactones appears to be
the usual result of the reaction of substituted acetylenes with allyl
halides (Table 7.11). Most notable is the general preference for
formation of five-membered lactones and ketones. Formation of cyclohexenones only becomes important in reactions with methallyl
chloride. It is not clear whether the phenols are formed from these
cyclohexenones or by cyclisation of a 2,5-dienoyl precursor. Examination of the structures of the products so far elucidated shows that

TABLE 7.11
ALLYL CHLORIDE

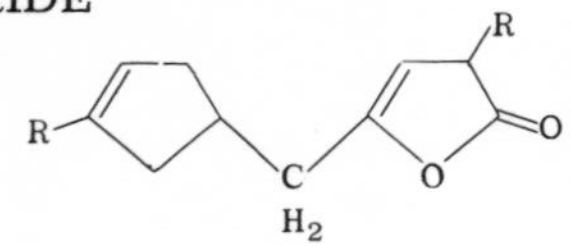

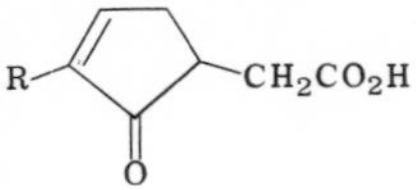

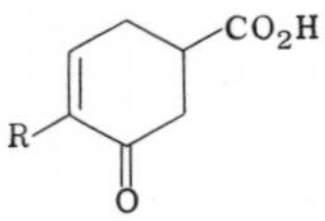

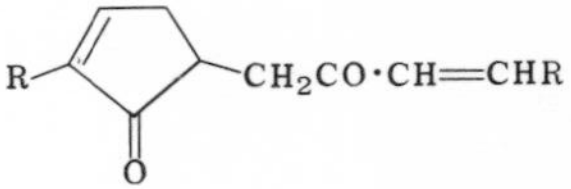

	A	B	C	D
Acetylene[24, 25]	R = H	—	—	—
Propyne[41, 42]	R = CH₃	—	R = CH₃	R = CH₃
Hex-1-yne[42]	R = C₄H₉(58%)	—	R = C₄H₉	—
Oct-1-yne[42]	R = C₆H₁₃(52%)	R = C₆H₁₃	R = C₆H₁₃	—
Phenylacetylene[42, 44, 45]	R = C₆H₅(4%)	R = C₆H₅(21%)	R = C₆H₅(4%)	R = C₆H₅(64%)

A: structure with R, CH_2
B: structure CH_2CO_2H
C: structure CO_2H
D: structure $CH_2CO \cdot CH=CHR$

CROTYL CHLORIDE

Hex-1-yne[45]

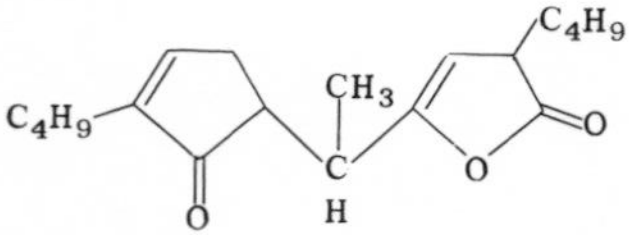

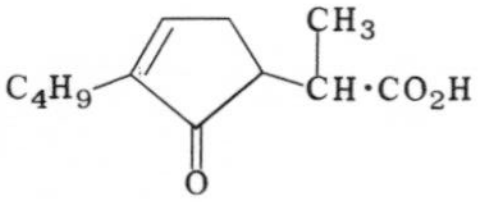

20%

Phenylacetylene[44, 45]

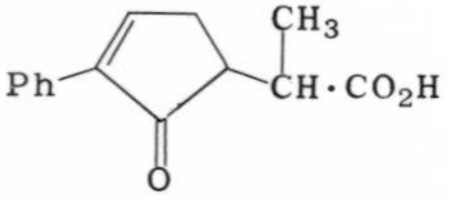

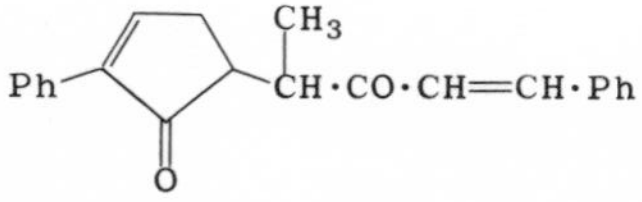

71%

(Continued overleaf)

METHYL 4-BROMOCROTONATE

Acetylene[24]

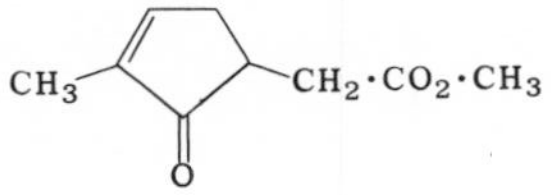

$CH_2 \cdot CO_2 \cdot CH_3$

CH_3

Phenylacetylene[45]

$CH_2 \cdot CO_2 \cdot CH_3$

Ph

Ph

CH_3
CO_2

Ph

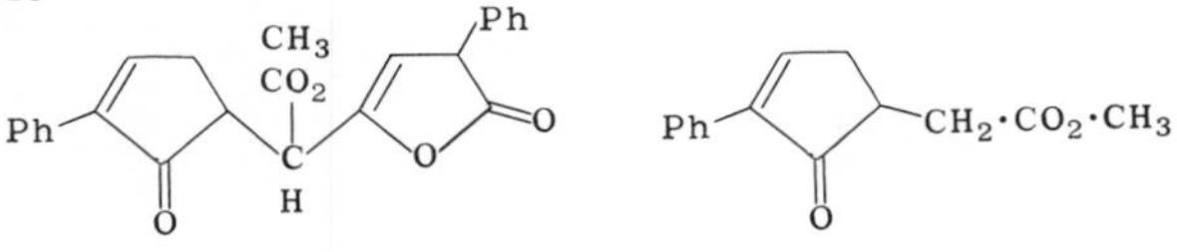

METHYL 4-BROMO-3-METHYLCROTONATE

Acetylene[25]

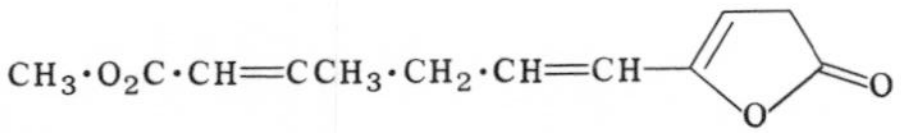

$CH_3 \cdot O_2C \cdot CH = CCH_3 \cdot CH_2 \cdot CH = CH$

METHALLYL CHLORIDE

Acetylene[25]

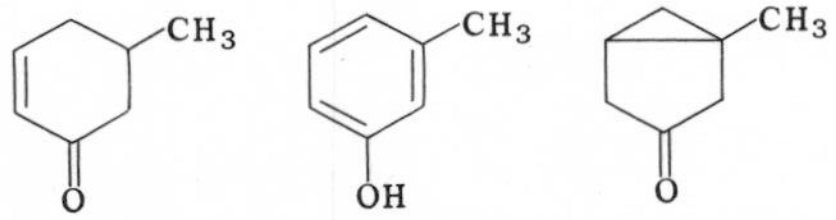

CH_3

CH_3

OH

CH_3

189

TABLE 7.11 (Contd.)

METHALLYL CHLORIDE (Contd.)

Hex-1-yne[45]

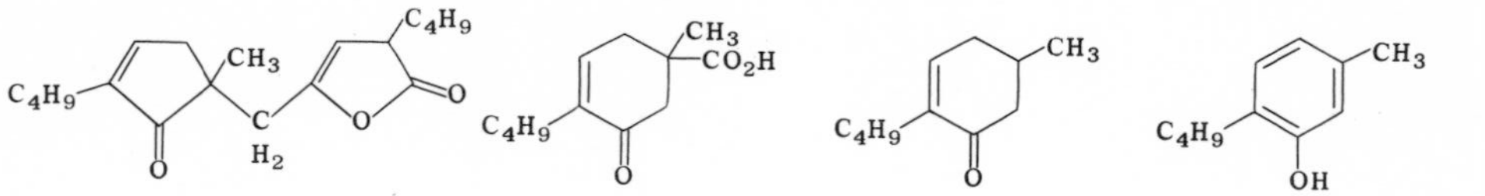

Phenylacetylene[44, 45]

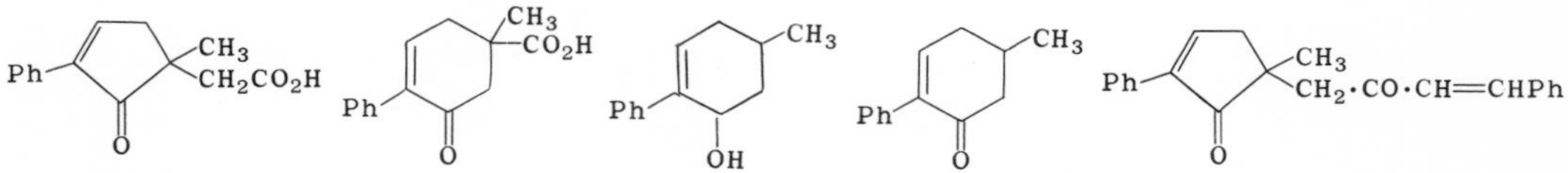

13% 34% 6% 37%

the allyl group becomes attached by its less encumbered end, to the end of the triple bond which acquires a hydrogen atom in simple hydrocarboxylation. A hypothetical scheme which depicts the origin of many of the products can be envisaged, (Figure 7. 5) in accordance with the known reactions of acyl-metal carbonyl species.

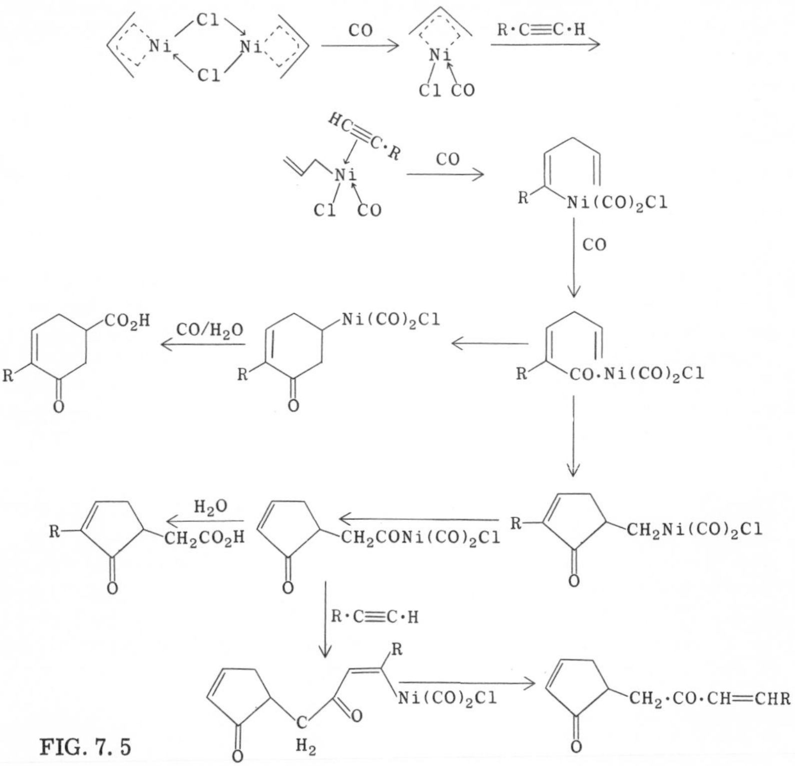

FIG. 7. 5

Under more vigorous conditions, 120-140°C and 30 atmospheres, iodobenzene reacts with acetylene and nickel carbonyl forming β-benzoylpropionic acid derivatives[55].

$$C_6H_5I + C_2H_2 + Ni(CO)_4 + ROH + HCl \longrightarrow$$

$$C_6H_5CO\ CH_2CH_2CO_2R + NiICl + 2CO$$

The anticipated β-benzoylacrylic compounds are reduced under the reaction conditions. An alternative mechanism is the reaction of $C_6H_5.CO.Ni(CO)_2I$ with preformed acrylic acid. Reactions of this type between allyl halides and acrylic acid derivatives have been observed[33].

$$CH_2 = CH.CH_2Cl + CH_2 = CH.CO_2H \xrightarrow{\text{Ni}} CH_2 = CH(CH_2)_3CO_2H$$

The difference in product type between aryl iodide and allyl halide probably results from the greater stability of the putative C_6H_5CO $Ni(CO)_2I$ intermediate over its but-3-enoyl analogue[55]. But-3-enoyl chloride is rapidly decarbonylated in the presence of nickel carbonyl and converted into hexa-1, 5-diene in the absence of substrate.

$$2CH_2{=}CH.CH_2CO\ Cl + Ni(CO)_4 \longrightarrow (CH_2CH{=}CH_2)_2 + NiCl_2 + 6CO.$$

The reactivity of iodobenzene in the foregoing reaction is mirrored in the reaction of aryl iodides with nickel carbonyl in refluxing methanol. Good yields of the methyl esters of benzoic acid (60%), m-chlorobenzoic acid (70%) and α-naphthoic acid (80%) have been obtained from the corresponding iodo compounds[10]. Aryl bromides and chlorides do not react under these conditions and necessitate more vigorous ones. Typical conditions employ nickel carbonyl, aqueous acid and carbon monoxide at 300°C and 600 atmospheres. The appropriate benzoic acid is obtained in about 30% yield (Table 7.12). Both iron[8, 18] and cobalt[8, 18, 179] carbonyls have also been used, and the use of a copper-zinc catalyst has been patented[107].

A variety of modifications to the general procedure have been employed. Thus, the corresponding esters are obtained if an anhydrous alcohol[106, 127] or lower ester[106, 153], such as methyl formate is used as the reaction medium. If methyl formate and sodium carbonate are used the pressurisation with carbon monoxide is unnecessary[104]. Amides or nitriles are produced in the presence of formamide, urea or oxamide[154]. The aroyl fluoride is obtained by the use of sodium fluoride under anhydrous conditions[126].

A further ramification of this reaction is the disproportionation of benzoic anhydrides to phthalic anhydrides and benzene, when heated with nickel carbonyl at 325°C and a carbon monoxide pressure of 100 atmospheres. Yields of 6 to 70% have been claimed[128, 129]. Similarly N, N-dibenzoylaniline disproportionates into N-phenylphthalimide and benzene[128]. It is possible to carry out both the carboxylation and disproportionation reactions concurrently if an alkali metal carbonate is present[129].

Arylcarboxylic acids are also obtained by the reaction of aryl diazonium salts with metal carbonyls at ambient pressure and temperature (Table 7.13). This route provides an acceptable single-step alternative to the Sandmeyer reaction especially in the presence of other hydrolytically labile groups[56, 144]. Both the chloride and tetrafluoroborate salts have been used with either nickel carbonyl, iron pentacarbonyl or dicobalt octacarbonyl. The hexacarbonyls of

TABLE 7.12 Preparation of Aryl Acids and Derivatives

Starting Material	Product	References
Chlorobenzene	Methyl benzoate	127, 137
Bromobenzene	Benzoic acid	104
Iodobenzene	Benzoic acid	179a
p-Methoxychlorobenzene	Anisic acid	104
p-Chlorotoluene	p-Toluic acid	18
p-Chlorotoluene	40% Toluonitrile and 40% p-Toluamide	154
p-Dichlorobenzene	p-Chlorobenzoic and terephthalic acids	8, 104, 127, 137, 179a
p-Chlorobenzoic acid	Terephthalic acid	104, 137
p-Dibromobenzene	p-Bromobenzoic and terephthalic acids	152
o-Dichlorobenzene	o-Chlorobenzoic and phthalic acids	179a
1-Chloronaphthalene	1-Naphthoic acid	179a
Benzene diazonium fluoroborate	Benzoyl fluoride	108
Chlorobenzene	Benzoyl fluoride	126
Bromobenzene	Benzoyl fluoride	126
o-Dichlorobenzene	o-Chlorobenzoyl fluoride	126
p-Dichlorobenzene	p-Chlorobenzoyl fluoride	126

TABLE 7.13 Carboxylation of Diazonium Salts

Diazonium Salt	Metal Carbonyl	Product	Yield (%)	Reference
Benzenediazonium Cl$^-$	Fe(CO)$_5$	Benzoic acid	11	144
p-Toluenediazonium BF$_4$$^-$	Ni(CO)$_4$	p-Toluic acid	19	56
Cl$^-$	Fe(CO)$_5$	p-Toluic acid	12	144
p-Chlorobenzenediazonium BF$_4$$^-$	Ni(CO)$_4$	p-Chlorobenzoic acid	39	56
Cl$^-$	Ni(CO)$_4$	p-Chlorobenzoic acid	—	144
Cl$^-$	Fe(CO)$_5$	p-Chlorobenzoic acid	41	144
Cl$^-$	Co$_2$(CO)$_8$	p-Chlorobenzoic acid	28	144
p-Methoxybenzenediazonium BF$_4$$^-$	Ni(CO)$_4$	p-Anisic acid	74	56
Cl$^-$	Fe(CO)$_5$	p-Anisic acid	17	144
Benzenediazonium-2-carboxylate	Ni(CO)$_4$	Phthalic anhydride	15–20	180
p-Ethoxycarbonylbenzenediazonium BF$_4$$^-$	Ni(CO)$_4$	Monoethyl terephthalate	52	56
p-Carboxybenzenediazonium BF$_4$$^-$	Ni(CO)$_4$	Terephthalic acid	76.5	56
BF$_4$$^-$	Fe(CO)$_5$	Terephthalic acid	14	56
p-Nitrobenzenediazonium Cl$^-$	Fe(CO)$_5$	p-Nitrobenzoic acid	25	144
1-Naphthalenediazonium BF$_4$$^-$	Ni(CO)$_4$	1-Naphthoic acid	4	56
Cl$^-$	Fe(CO)$_5$	1-Naphthoic acid	trace	144

molybdenum and chromium are unreactive. The use of aqueous solvents gives the acid, and ethanol the ethyl ester often accompanied by reduction products of the diazonium salt and small amounts of diarylketone. The reaction of benzenediazonium tetrafluoroborate in in anhydrous alcohol saturated with hydrogen chloride yielded benzoyl fluoride[108].

Extension of the reaction to diazoalkanes shows that in the presence of a small amount of nickel carbonyl only substituted ethylenes and ketazines are generated[141]. However, good yields of diphenylketen and diphenyleneketen are obtained from the reaction of diphenyl-

$$Ph_2CN_2 \longrightarrow Ph_2C = C \, Ph_2 + Ph_2C = N.N = C \, Ph_2$$

diazomethane and 9-diazofluorene with an excess of nickel carbonyl.

$$Ph_2C.N_2 \xrightarrow{Ni(CO)_4} Ph_2C = C = 0 + N_2$$

Similarly in ethanolic solution ethyl diazoacetate yields some diethyl malonate.

REFERENCES

1. Adkins, H. and Rosenthal, R. W., J. Amer. Chem. Soc., 1950, **72**, 4550.

2. Ajinomoto Co., Inc., Neth. Appl. 6, 413, 065; Chem. Abs., 1965, **63**, 16219.

3. Albanesi, G. and Gavezzotti, E., Chimica e Industria, 1965, **47**, 1322.

4. Alderson, T. and Engelhardt, V. A., U.S.P. 3, 065, 242; Chem. Abs., 1963, **58**, 8912.

5. Aliev, Y. Y. and Isakov, Y. I., Issled. Mineral'n. i Rast Syr' ya Uzbekistana, Akad. Nauk Uz. S.S.R., Inst. Khim., 1962, 95; Chem. Abs., 1963, **59**, 3767.

6. Aliev, Y. Y. and Romanova, I. B., Neftekhim., Akad. Nauk Turk. S.S.R., 1963, 204; Chem. Abs., 1964, **61**, 6913.

7. Ashworth, P. J., Whitham, G. H. and Whiting, M. C., J. Chem. Soc., 1957, 4633.

8. Badische Anilin-und Soda-Fabrik, B.P. 815, 835; Chem. Abs., 1960, **54**, 1449.

9. Badische Anilin-und Soda-Fabrik, Neth. Appl. 6, 409, 121; Chem. Abs., 1965, **63**, 14726.

10. Bauld, N. L., Tetrahedron Letters, 1963, 1841.

11. Benson, R. E., U.S.P. 2, 871, 262; Chem. Abs., 1959, 53, 14008.

12. Bergman, E. D. and Zimkin, E., J. Chem. Soc., 1950, 3455.

13. Bhattacharyya, S. K. and Sen, A. K., J. Appl. Chem., 1963, 13, 498.

14. Bird, C. W., Cookson, R. C., Hudec, J. and Williams, R. O., J. Chem. Soc., 1963, 410.

15. Bird, C. W. and Hollins, E. M., unpublished work.

16. Bird, C. W. and Hudec, J., Chem. and Ind., 1959, 570.

17. Blackham, A. U., U.S.P. 3, 119, 861; Chem. Abs., 1964, 60, 9155.

18. Bliss, H. and Southworth, R. W., U.S.P. 2, 565, 461; Chem. Abs., 1952, 46, 2577.

19. Brewis, S. and Hughes, P. R., Chem. Comm., 1965, 157.

20. Brewis, S. and Hughes, P. R., Chem. Comm., 1965, 489.

21. Brewis, S. and Hughes, P. R., Chem. Comm., 1966, 6.

22. Brooks, R. E., U.S.P. 2, 457, 204; Chem. Abs., 1949, 43, 3443.

23. Case, J. R., Clarkson, R., Jones, E. R. H. and Whiting, M. C., Proc. Chem. Soc., 1959, 150.

24. Cassar, L. and Chiusoli, G. P., Tetrahedron Letters, 1965, 3295.

25. Cassar, L. and Chiusoli, G. P., Chimica e Industria, 1966, 48, 323.

26. Cassar, L. and Chiusoli, G. P., Tetrahedron Letters, 1966, 2805.

27. Chiusoli, G. P., Chimica e Industria, 1959, 41, 503.

28. Chiusoli, G. P., Chimica e Industria, 1959, 41, 506.

29. Chiusoli, G. P., Chimica e Industria, 1959, 41, 512.

30. Chiusoli, G. P., Chimica e Industria, 1959, 41, 762.

31. Chiusoli, G. P., Gazzetta, 1959, 89, 1332.

32. Chiusoli, G. P., Angew. Chem., 1960, 72, 74.

33. Chiusoli, G. P., Chimica e Industria, 1961, 43, 365.

34. Chiusoli, G. P., U.S.P. 3, 025, 320; Chem. Abs., 1962, 57, 9668.

35. Chiusoli, G. P., U.S.P. 3, 146, 257; Chem. Abs., 1965, 62, 9015.

36. Chiusoli, G. P. and Agnes, G., Z. Naturforsch., 1962, 17b, 852.

37. Chiusoli, G. P. and Agnes, G., Proc. Chem. Soc., 1963, 310.

38. Chiusoli, G. P. and Agnes, G., Chimica e Industria, 1964, 46, 548.

39. Chiusoli, G. P. and Agnes, G., Chimica e Industria, 1964, **46**, 548-9.

40. Chiusoli, G. P. and Bottaccio, G., Chimica e Industria, 1961, **43**, 1022.

41. Chiusoli, G. P. and Bottaccio, G., Chimica e Industria, 1962, **44**, 1129.

42. Chiusoli, G. P. and Bottaccio, G., Chimica e Industria, 1965, **47**, 165.

43. Chiusoli, G. P., Bottaccio, G. and Cameroni, A., Chimica e Industria, 1962, **44**, 131.

44. Chiusoli, G. P., Bottaccio, G. and Venturello, C., Tetrahedron Letters, 1965, 2875.

45. Chiusoli, G. P., Bottaccio, G. and Venturello, C., Chimica e Industria, 1966, **48**, 107.

46. Chiusoli, G. P. and Cameroni, A., Chimica e Industria, 1964, **46**, 1063.

47. Chiusoli, G. P. and Merzoni, S., Chimica e Industria, 1961, **43**, 259.

48. Chiusoli, G. P. and Merzoni, S., U.S.P. 3, 032, 583; Chem. Abs., 1962, **57**, 11030.

49. Chiusoli, G. P. and Merzoni, S., Z. Naturforsch, 1962, **17b**, 850.

50. Chiusoli, G. P. and Merzoni, S., Belg. P. 618, 557; Chem. Abs., 1963, **59**, 11268.

51. Chiusoli, G. P. and Merzoni, S., Chimica e Industria, 1963, **45**, 6.

52. Chiusoli, G. P. and Merzoni, S., Fr. P. 1, 342, 549; Chem. Abs., 1964, **60**, 13146.

53. Chiusoli, G. P. and Merzoni, S., Ital. P. 675, 616; Chem. Abs., 1966, **64**, 3363.

54. Chiusoli, G. P., Merzoni, S. and Mondelli, G., Chimica e Industria, 1964, **46**, 743.

55. Chiusoli, G. P., Merzoni, S. and Mondelli, G., Tetrahedron Letters, 1964, 2777.

56. Clark, J. C. and Cookson, R. C., J. Chem. Soc., 1962, 686.

57. Clarkson, R., Jones, E. R. H., Wailes, P. C. and Whiting, M. C., J. Amer. Chem. Soc., 1956, **78**, 6206.

58. Codignola, F. and Piacenza, M., Ital. P. 431, 407; Chem. Abs., 1950, **44**, 1134.

59. Crowe, B. F., Chem. and Ind., 1960, 1000.

60. Crowe, B. F., Chem. and Ind., 1960, 1506.

61. Culvenor, C. C. J. and Geissman, T. A., Chem. and Ind., 1959, 366.

62. Culvenor, C. C. J. and Geissman, T. A., J. Amer. Chem. Soc., 1961, 83, 1647.

63. Dent, W. T., Long, R. and Whitfield, G. H., J. Chem. Soc., 1964, 1588.

64. Dubini, M., Chiusoli, G. P. and Montino, F., Tetrahedron Letters, 1963, 1591.

65. Dunn, J. T., B.P. 879, 009; 879, 010; 879, 052; 879, 305; 879, 306. Chem. Abs., 1963, 59, 1494 ff.

66. Du Pont, G., Pignaniol, P., and Vialle, J., Bull. Soc. chim. France, 1948, 529.

67. Ehrreich, J. E., Nickerson, R. G. and Ziegler, C. E., Ind. and Eng. Chem. (Process Design), 1965, 4, 77.

68. Eisenmann, J. L., J. Org. Chem., 1962, 27, 2706.

69. Eisenmann, J. L., Yamartino, R. L. and Howard, J. F., J. Org. Chem., 1961, 26, 2102.

70. Ercoli, R., U.S.P. 2, 911, 422; Chem. Abs., 1960, 54, 7590.

71. Ercoli, R., Signorini, G. and Santambrogio, E., Chimica e Industria, 1960, 42, 587.

72. Falbe, J. and Korte, F., Angew. Chem. Internat. Edn., 1962, 1, 266.

73. Falbe, J. and Korte, F., Chem. Ber., 1962, 95, 2680.

74. Falbe, J. and Korte, F., Chem. Ber., 1965, 98, 1928.

75. Falbe, J., Schulze-Steinen, H. J. and Korte, F., Chem. Ber., 1965, 98, 886.

76. Falbe, J., Weitkamp, H. and Korte, F., Tetrahedron Letters, 1965, 2677.

77. Fell, B. and Tetteroo, J. M. J., Angew. Chem. Internat. Edn., 1965, 4, 790

78. Furumi, M., Mizoroki, T., Nakayama, M. and Ando, Y., Reports Govt. Chem. Ind. Res. Inst., Tokyo, 1963, 58, 293. Chem. Abs., 1965, 62, 6388.

79. Genas, M. and Rull, T., Fr.P. 1, 383, 726; Chem. Abs., 1965, 62, 9034.

80. Gresham, W. F., U.S.P. 2, 432, 474; Chem. Abs., 1948, 42, 1961.

81. Gresham, W. F. and Brooks, R. E., U.S.P. 2, 448, 368; Chem. Abs., 1949, **43**, 669.

82. Gresham, W. F. and Brooks, R. E., B.P. 631, 001; Chem. Abs., 1950, **44**, 4493. U.S.P. 2, 549, 453; Chem. Abs., 1951, **45**, 8551.

83. Hagemeyer, H. J., U.S.P. 2, 739, 169; Chem. Abs., 1956, **50**, 16835

84. Happel, J. and Marsell, C. J., Belg. P. 639, 260; Chem. Abs., 1965, **62**, 9019.

85. Heck, R. F., J. Amer. Chem. Soc., 1963, **85**, 1460.

86. Heck, R. F., J. Amer. Chem. Soc., 1963, **85**, 2013.

87. Heck, R. F., U.S.P. 3, 116, 306; Chem. Abs., 1964, **60**, 6794.

88. Heck, R. F. and Breslow, D. S., J. Amer. Chem. Soc., 1963, **85**, 2779.

89. Himmele, W. and Kutepow, N. V., B.P. 899, 691; Chem. Abs., 1963, **59**, 1491.

90. Hines, R. A., U.S.P. 2, 809, 991; Chem. Abs., 1958, **52**, 3856.

91. Hock, A. A. and Mills, O. S., Proc. Chem. Soc., 1958, 233.

92. Inoue, T. and Tsutsumi, S., Bull. Chem. Soc., Japan, 1965, **38**, 2122.

93. Inoue, T. and Tsutsumi, S., J. Amer. Chem. Soc., 1965, **87**, 3525.

94. Jacobsen, G. and Spathe, H., G.P. 1, 138, 760; Chem. Abs., 1963, **58**, 6699.

95. Jenner, E. L. and Lindsey, R. V., U.S.P. 2, 876, 254; Chem. Abs., 1959, **53**, 17906.

96. Jones, E. R. H., Shen, T. Y. and Whiting, M. C., J. Chem. Soc. 1950, 230.

97. Jones, E. R. H., Shen, T. Y. and Whiting, M. C., J. Chem. Soc., 1951, 48.

98. Jones, E. R. H., Shen, T. Y. and Whiting, M. C., J. Chem. Soc., 1951, 763.

99. Jones, E. R. H., Shen, T. Y. and Whiting, M. C., J. Chem. Soc., 1951, 766.

100. Jones, E. R. H., Whitham, G. H. and Whiting, M. C., J. Chem. Soc., 1954, 1865.

101. Jones, E. R. H., Whitham, G. H. and Whiting, M. C., J. Chem. Soc., 1957, 4628.

102. Kealy, T. J. and Benson, R. E., J. Org. Chem., 1961, **26**, 3126.

103. Klemchuk, P. P., U.S.P. 3, 064, 040; Chem. Abs., 1963, **58**, 11221.

104. Kröper, H., Wirth, F. and Huckler, O. H., G.P. 1, 074, 028; Chem. Abs., 1961, **55**, 12364.

105. Larson, A. T., U.S.P. 2, 448, 375; Chem. Abs., 1949, **43**, 670.

106. Leibeu, H. J., U.S.P. 2, 640, 071; Chem. Abs., 1954, **48**, 5214.

107. Leibeu, H. J., U.S.P. 2, 691, 671; Chem. Abs., 1955, **49**, 14807.

108. Linville, R. G., U.S.P. 2, 517, 898; Chem. Abs., 1951, **45**, 2505.

109. Long, R. and Whitfield, G. H., B.P. 983, 341; Chem. Abs., 1965, **62**, 16065.

110. Long, R. and Whitfield, G. H., B.P. 1, 007, 707; Chem. Abs., 1966, **64**, 3361.

111. Matsuda, A. and Uchida, H., Bull. Chem. Soc. Japan, 1965, **38**, 710.

112. Mizoroki, T. and Nakayama, M., Bull. Chem. Soc. Japan, 1965, **38**, 1876.

113. Mizoroki, T., Nakayama, M., Ando, Y. and Furumi, M., J. Chem. Soc. Japan, Ind. Chem. Sect., 1962, **65**, 1049.

114. Mizoroki, T., Nakayama, M. and Furumi, M., J. Chem. Soc. Japan, Ind. Chem. Sect., 1962, **65**, 1054.

115. "Montecatini" Societa Generale per l'Industria Mineraria e Chimica, B.P. 888, 162. Chem. Abs., 1964, **61**, 8195.

116. Mueller, G. P. and MacArtor, F. L., J. Amer. Chem. Soc., 1954, **76**, 4621.

117. Natta, G. and Pino, P., Chimica e Industria, 1952, **34**, 449.

118. Newitt, D. M. and Momen, S. A., J. Chem. Soc., 1949, 2945.

119. Piacenti, F. and Cioni, C., Atti Soc. Toscana Sci. Nat. Pisa, Proc. Verbali Mem., 1962, **69b**, 1; Chem. Abs., 1965, **63**, 11347.

120. Piacenti, F., Neggiani, P.D. and Calderazzo, F., Atti. Soc. Toscana Sci. Nat. Pisa, Proc. Verbali Mem., 1962, **69b**, 42; Chem. Abs., 1965, **63**, 11347.

121. Pino, P. and Magri, R., Chimica e Industria, 1952, **34**, 511.

122. Pino, P. and Miglierina, A., J. Amer. Chem. Soc., 1952, **74**, 5551.

123. Pino, P., Miglierina, A. and Pietra, E., Gazzetta, 1954, **84**, 443.

124. Pino, P. and Paleari, C., Gazzetta, 1951, **81**, 646.

125. Pino, P., Pietra, E. and Mondello, B., Gazzetta, 1954, **84**, 453.

126. Prichard, W. W., U.S.P. 2, 696, 503; Chem. Abs., 1955, **49**, 15966.

127. Prichard, W. W. and Tabet, G. E., U.S.P. 2, 565, 462; Chem. Abs., 1952, **46**, 2578.

128. Prichard, W. W., U.S.P. 2, 680, 750; Chem. Abs., 1955, **49**, 6308.

129. Prichard, W. W., U.S.P. 2, 680, 751; Chem. Abs., 1955, **49**, 6308.

130. Reppe, W., Annalen, 1953, **582**, 1.

131. Reppe, W. and Kroper, H., G.P. 765, 969; Chem. Abs., 1957, **51**, 13904.

132. Reppe, W. and Kroper, H., G.P. 863, 194; Chem. Abs., 1954, **48**, 1425.

133. Reppe, W. and Kroper, H., Annalen, 1953, **582**, 38.

134. Reppe, W., Kroper, H., Kutepow, N. and Pistor, H. J., Annalen, 1953, **582**, 72.

135. Reppe, W., Kroper, H., Pistor, H. J. and Weissbarth, O., Annalen, 1953, **582**, 87.

135a. Reppe, W. and Magin, A., U.S.P. 2, 577, 208; Chem. Abs., 1952, **46**, 6143.

136. Reppe, W., Schlichting, O., Klager, K. and Toepel, T., Annalen, 1948, **560**, 1.

137. Romanovskii, V. I. and Artem'ev, A. A., Zhur. Vsesoyuz. Khim. obshch. im. D.I. Mendeleeva, 1960, **5**, 476.

138. Rosenthal, R. W., U.S.P. 2, 652, 413; Chem. Abs., 1954, **48**, 5209.

139. Rosenthal, R. W. and Schwartzman, L. H., J. Org. Chem., 1959, **24**, 836.

140. Rosenthal, R. W., Schwartzman, L. H., Grego, N. P. and Proper, R., J. Org. Chem., 1963, **28**, 2835.

141. Rüchardt, C. and Schrauzer, G. N., Chem. Ber., 1960, **93**, 1840.

142. Ruell, T., Bull. Soc. chim. France, 1964, 2680.

143. Sakakibara, Y., Bull. Chem. Soc. Japan, 1964, **37**, 1601.

144. Schrauzer, G. N., Chem. Ber., 1961, **94**, 1891.

145. Schwartzman, L. H. and Rosenthal, R. W., Abs. A.C.S. Meeting, April 1959, p. 57-O

146. Seon, M. and Leleu, J., U.S.P. 2, 782, 226; Chem. Abs., 1957, **51**, 10564.

147. Shell Internationale Research Maatschappij N.V., Neth. Appl., 6, 408, 476; Chem. Abs., 1965, **63**, 499.

148. Sternberg, H. W., Friedel, R. A., Markby, R. and Wender, I., J. Amer. Chem. Soc., 1956, **78**, 3621.

149. Sternberg, H. W., Markby, R. and Wender, I., J. Amer. Chem. Soc., 1960, **82**, 3638.

150. Sternberg, H. W., Shukys, J. G., Donne, C. D., Markby, R., Friedel, R. A. and Wender, I., J. Amer. Chem. Soc., 1959, **81**, 2339.

151. Striegler, A. and Weber, J., J. prakt. Chem., 1965, **29**, 281.

152. Tabet, G. E., U.S.P. 2, 565, 463; Chem. Abs., 1952, **46**, 2578.

153. Tabet, G. E., U.S.P. 2, 565, 464; Chem. Abs., 1952, **46**, 2578.

154. Tabet, G. E., U.S.P. 2, 691, 670; Chem. Abs., 1955, **49**, 14806.

155. Takegami, Y., Yokokawa, C. and Watanabe, Y., Bull. Chem. Soc. Japan., 1964, **37**, 935.

156. Takegami, Y., Yokokawa, C., Watanabe, Y. and Masada, H., Bull. Chem. Soc. Japan, 1964, **37**, 672.

157. Takegami, Y., Yokokawa, C., Watanabe, Y. and Masada, H., Bull. Chem. Soc. Japan, 1965, **38**, 1649.

158. Toyo Rayon Co. Ltd., Fr.P. 1, 389, 856; Chem. Abs., 1965, **63**, 501.

159. Tsuji, J. and Hosaka, S., Jap. P. 5699('65); Chem. Abs., 1965, **62**, 16085.

160. Tsuji, J. and Hosaka, S., J. Amer. Chem. Soc., 1965, **87**, 4075.

161. Tsuji, J., Hosaka, S., Kiji, J. and Susuki, T., Bull. Chem. Soc. Japan, 1966, **39**, 141.

162. Tsuji, J., Imamura, S. and Kiji, J., J. Amer. Chem. Soc., 1964, **86**, 4491.

163. Tsuji, J., Kiji, J. and Hosaka, S., Tetrahedron Letters., 1964, 605.

164. Tsuji, J., Kiji, J., Imamura, S. and Morikawa, M., J. Amer. Chem. Soc., 1964, **86**, 4350.

165. Tsuji, J., Kiji, J. and Morikawa, M., Tetrahedron Letters, 1963, 1811.

166. Tsuji, J., Morikawa, M. and Iwamoto, N., J. Amer. Chem. Soc., 1964, **86**, 2095.

167. Tsuji, J., Morikawa, M. and Kiji, J., Tetrahedron Letters, 1963, 1061.

168. Tsuji, J., Morikawa, M. and Kiji, J., Tetrahedron Letters, 1963, 1437.

169. Tsuji, J., Morikawa, M. and Kiji, J., J. Amer. Chem. Soc., 1964, **86**, 4851.

170. Tsuji, J., Morikawa, M. and Kiji, J., Tetrahedron Letters, 1965, 817.

171. Tsuji, J., Morikawa, M. and Kiji, J., Jap. P. 19, 940('65); Chem. Abs., 1965, 63, 16219.

172. Tsuji, J. and Nogi, T., Bull. Chem. Soc. Japan, 1966, 39, 146.

173. Tsuji, J. and Nogi, T., J. Amer. Chem. Soc., 1966, 88, 1289.

174. Tsuji, J. and Nogi, T., Tetrahedron Letters, 1966, 1801.

175. Tsuji, J. and Susuki, T., Tetrahedron Letters, 1965, 3027.

176. Weiss, E., Hübel, W. and Merenyi, R., Chem. Ber., 1962, 95, 1155.

177. Yakubovich, A. Y. and Volkova, E. V., Doklady Akad. Nauk S.S.S.R., 1952, 84, 1183.

178. Yakubovich, A. Y. and Volkova, E. V., Zhur. obschei Khim., 1960, 30, 3972.

179. Yamamoto, K., Bull. Chem. Soc. Japan, 1954, 27, 491.

179a. Yamamoto, K. and Sato, K., Jap. P. 2424('52); Chem. Abs., 1954, 48, 2105.

180. Yaroslavsky, S., Chem. and Ind., 1965, 765.

181. Zachry, J. B. and Aldridge, C. L., U.S.P. 3, 161, 672; Chem. Abs., 1965, 62, 9018.

182. Zachry, J. B. and Aldridge, C. L., U.S.P. 3, 176, 038; Chem. Abs., 1965, 62, 16064.

CHAPTER 8

OTHER CARBONYLATION REACTIONS

The main concern of this chapter is reactions which formally entail the incorporation of a molecule of carbon monoxide. Although the term "carbonylation" would be confined ideally to reactions of this type it is now used widely to describe all reactions entailing incorporation of carbon monoxide.

As in the case of the hydroformylation and hydrocarboxylation reactions acyl-metal carbonyl species are the key intermediates. This is illustrated by the reaction of but-1-ene with cobalt hydrocarbonyl.

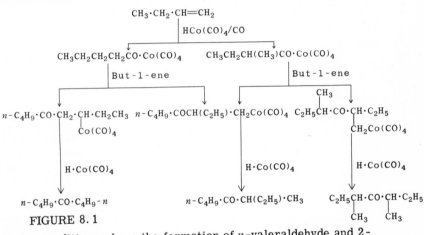

FIGURE 8.1

Under conditions where the formation of n-valeraldehyde and 2-methylbutyraldehyde is minimised, the main products are nonan-5-one, 3-methyloctan-4-one and 3,5-dimethylheptan-4-one[14], as a result of addition of pentanoylcobalt tetracarbonyl to but-1-ene, (Figure 8.1) followed by reduction. An intramolecular version can be effected with alkenoylcobalt tetracarbonyls[36]. These compounds are prepared at 0°C from the alkenoyl halide and sodium tetracarbonylcobaltate and decompose at higher temperatures. Cyclic ketones are the major products with hex-5-enoyl or in lesser amounts from hept-6-enoyl derivatives. In other cases π-allyl or π-alkenoyl cobalt tricarbonyl complexes result. As in the case of (1) five-membered ring formation predominates. The 2-methylcyclopent-2-enone presumably arises from isomerisation of 2-methylenecyclopentanone. The hept-6-enoylcobalt tetracarbonyl gives only 2-

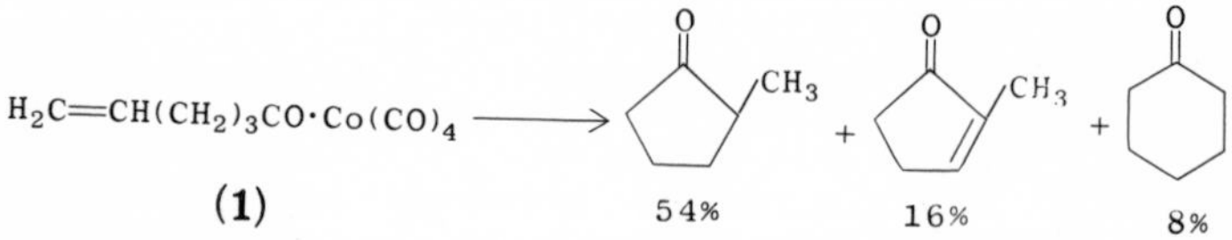

$$H_2C=CH(CH_2)_3CO \cdot Co(CO)_4 \longrightarrow$$

(1)

54% 16% 8%

methylcyclohexanone; no cycloheptanone is detectable. Non-conjugated dienes can be converted catalytically into cyclopentanones by reaction with dicobalt octacarbonyl and carbon monoxide in aqueous acetone at 100-200°C and 70-340 atmospheres[58].

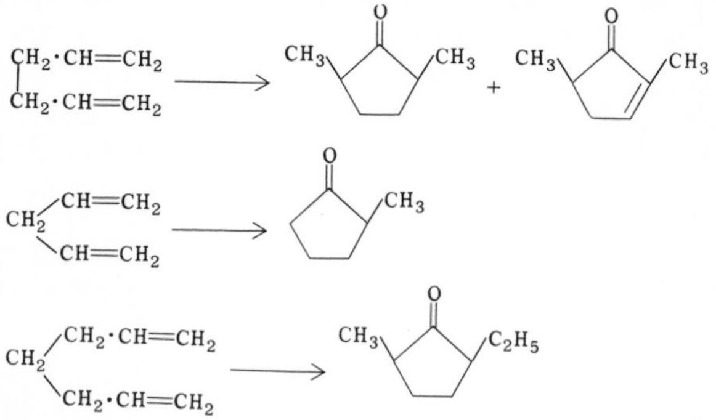

Penta-1,3-diene, which can isomerise to the 1,4-diene, yields 2-methylcyclopentanone and cyclohexanone with cobalt hydrocarbonyl[14]. A synthetic route to 3-hydroxyfurans is suggested by the reaction of allyloxyacetylcobalt tetracarbonyl (2) with dicyclohexylethylamine[36].

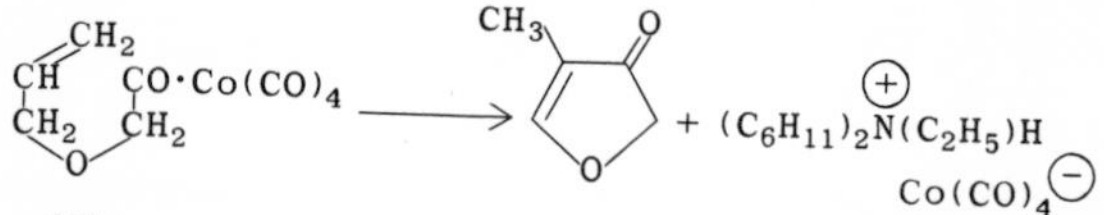

(2)

The reaction of alkyl or acyl cobalt tetracarbonyls with conjugated dienes gives 1-acylmethyl-π-allylcobalt tricarbonyls[37,38,41], and these on treatment with dicyclohexylethylamine or sodium methoxide produce 1-acylbuta-1,3-dienes (Table 8.1). The reaction can be made catalytic with respect to tetracarbonylcobaltate anion, the

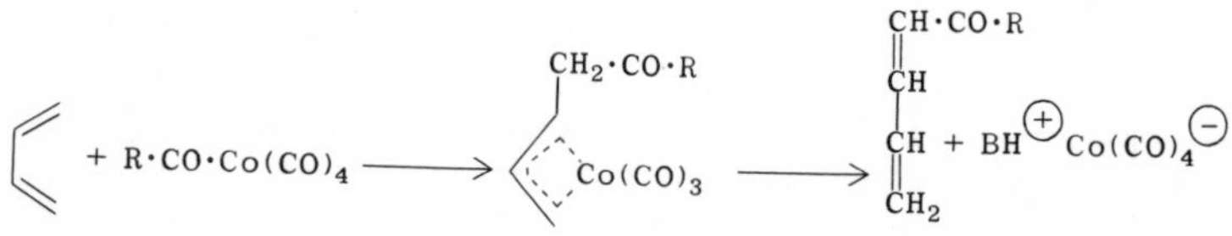

206

TABLE 8.1 Preparation of 1-Acylbuta-1,3-dienes[38,41]

Diene	Alkyl or Acyl Halide	Product
Butadiene	Methyl iodide	Hexa-1,3-dien-5-one (20%)
	Ethyl iodide	Hepta-1,3-dien-5-one
	"Crotyl bromide"	Nona-1,3,7-trien-5-one
	Chlorodimethyl ether	1-Methoxyhexa-3,5-dien-2-one
	Methyl chloroacetate	Methyl 3-ketohepta-4,6-dienoate
	Benzoyl bromide	1-Benzoylbuta-1,3-diene
trans-Penta-1,3-diene	Methyl iodide	Hepta-2,4-dien-6-one
cis-Penta-1,3-diene	Methyl iodide	3-Methylhexa-3,5-dien-2-one (15%)
Isoprene	Methyl iodide	2-Methylhexa-1,3-dien-5-one (10%)
	Cyclohexene oxide	1-(2-Hydroxycyclohexyl)-5-methylpenta-2,4-dien-1-one
	Pivalyl chloride	2,2,6-Trimethylhepta-4,6-dien-3-one
	Acrylyl chloride	2-Methylhepta-1,3,6-trien-5-one
Cyclopentadiene	Methyl iodide	1-Acetylcyclopenta-1,3-diene
4-Methylpenta-1,3-diene	Methyl iodide	2-Methylhepta-2,4-dien-6-one
2,3-Dimethylbutadiene	Methyl iodide	2,3-Dimethylhexa-1,3-dien-5-one

(Continued overleaf)

207

TABLE 8.1 (continued)

Diene	Alkyl or Acyl Halide	Product
Cyclohexa-1,3-diene	Methyl iodide	1-Acetylcyclohexa-1,3-diene
	Diethyl sulphate	1-Propionylcyclohexa-1,3-diene
Myrcene	Methyl iodide	Acetylmyrcene
trans-1-Methoxybutadiene	Methyl iodide	1-Methoxyhexa-1,3-dien-5-one
	Chlorodimethyl ether	1,6-Dimethoxyhexa-1,3-dien-5-one
	Ethyl bromoacetate	Ethyl 3-keto-7-methoxyhepta-4,6-dienoate
Methyl sorbate	Methyl iodide	Methyl 6-keto-5-methylhepta-2,4-dienoate

other reactants being the diene, carbon monoxide and an alkyl or acyl halide. A reaction analogous to the first step in the above sequence has been observed in the reaction between butadiene and methyl or phenyl manganese pentacarbonyl[9,10]. In this case a 1,4-hydride shift occurs[33a].

(3)

The complex (3) is not converted to 1-acylbuta-1,3-diene by treatment with sodium methoxide. A different route to this type of compound is provided by the Friedel-Crafts acetylation of the butadiene iron tricarbonyl complex[31]. Decomposition of the mixture of acetylbutadiene iron tricarbonyl complexes with carbon monoxide at 175°C liberates a 4:1 mixture of 1- and 2-acetylbuta-1,3-diene.

The formation of cyclopentanones from olefins and carbon monoxide in the presence of a metal carbonyl has not been extensively studied. Cyclohexene reacts with carbon monoxide and hydrogen in a stainless steel autoclave at 300°C and 800 atmospheres pressure. The products isolated are cyclohexanecarboxaldehyde, cyclohexylcarbinol and perhydrofluorenone[60].

Perhydrofluorenone was also formed when a platinum lined autoclave was employed at 350°C. Despite the original conclusion that this excluded metal carbonyl catalysis it is now known that definite carbonyl complexes are formed on platinum surfaces[32], and many other examples of platinum catalysed carbonylation reactions are known. Cyclopentene under similar conditions gives the ketone (4).

(4)

The highly reactive bicyclo[2,2,1]heptadiene reacts with iron carbonyls under much milder conditions forming the ketones (5), (6) and (7), in addition to hydrocarbon dimers and the bicycloheptadiene iron

(5)

(6)

(7)

tricarbonyl complex[15,79]. The stereochemistry of (5) is now con-
clusively settled as *exo, trans, exo*.[28,34]. Bicyclo[2, 2, 1]heptene
does not react with iron pentacarbonyl under comparable conditions,
but with di-iron nonacarbonyl at room temperature it gives reasonable yields of tetrahydro (5). [15] Bicyclo[2, 2, 1]heptadiene under the
latter conditions gives mostly hydrocarbon dimers and small
amounts of (5). Benzobicyclo[2, 2, 1]heptadienes react more sluggishly with iron pentacarbonyl forming dibenzo-analogues of (5). The
reaction of bicyclo[2, 2, 1]heptene and its derivatives with carbon
monoxide, catalysed by dicobalt octacarbonyl, takes a different
course yielding the lactone (8a) as the principal product[27]. The
reaction can be further generalised to the formation of (8b) by carbonylation of bicyclo[2, 2, 1]heptenes with other olefins[85].

(8a)

(8b)

The reactions of acetylenes with iron carbonyls show many analogies
to the foregoing bicyclo[2, 2, 1]heptadiene reactions, and have been a
particularly fruitful source of novel organic and organometallic compounds. In addition to tetraphenylcyclopentadienone and its iron tricarbonyl complex the most important products formed from diphenylacetylene and iron carbonyls are (9), (10), (11) and (12) in the present context[23,53,107,115]. Essentially the same products are formed from diphenylcyclopropenone and iron carbonyls[16,17]. Complexes
(10) and (11) are converted into tetraphenylcyclopentadienone or its
iron tricarbonyl complex by irradiation, heating with carbon monox-

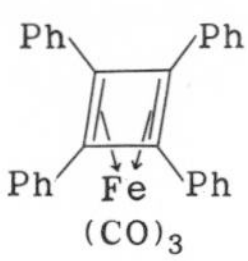

(9) 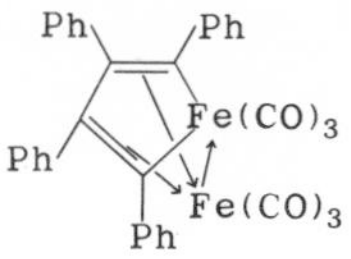   **(10)**

ide, reaction with bromine in acetic acid, or reduction of (10) with lithium aluminium hydride.

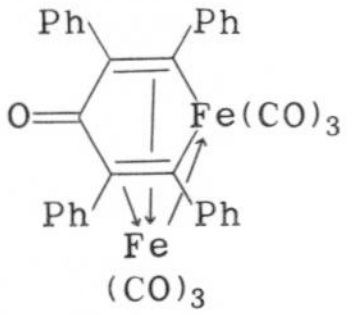

(11) 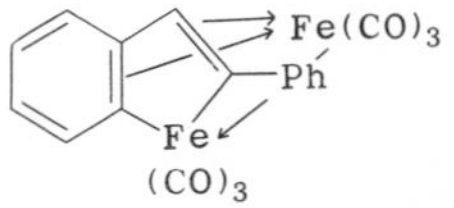   **(12)**

Reduction of complex (11) with either lithium aluminium hydride, sodium in liquid ammonia, or stannous chloride gives 2, 3, 5, 6-tetraphenylhydroquinone[53] whereas oxidation with nitric acid produces the quinone. The complex (11) has also been converted into 2, 3, 5, 6-tetraphenyl derivatives of 4-pyrone, thiapyrone and selen-

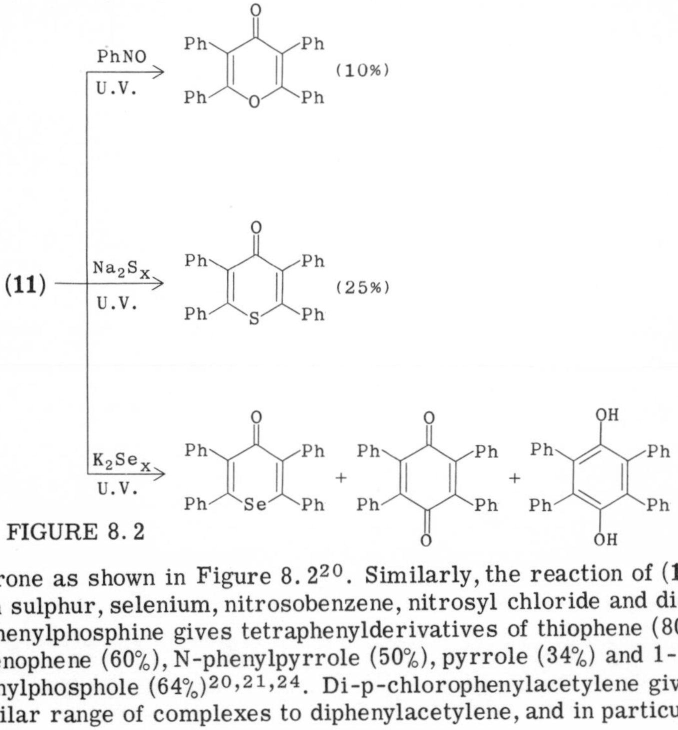

FIGURE 8. 2

apyrone as shown in Figure 8. 2[20]. Similarly, the reaction of (10) with sulphur, selenium, nitrosobenzene, nitrosyl chloride and dichlorophenylphosphine gives tetraphenylderivatives of thiophene (80%), selenophene (60%), N-phenylpyrrole (50%), pyrrole (34%) and 1-phenylphosphole (64%)[20,21,24]. Di-p-chlorophenylacetylene gives a similar range of complexes to diphenylacetylene, and in particular

211

(13) and (14)[23]. Treatment of (14) with carbon monoxide yields 2-(p-chlorophenyl)-6-chloroindenone.

(13)   (14)

Phenylacetylene gives a somewhate different range of products which include complexes (15) to (18) inclusive[22,53,56,61,106]. It is noteworthy that the 2, 5-diphenylcyclopentadienone complex (15) is formed in much larger amounts than its 3, 4-isomer (16), paralleling the behaviour of phenylacetylene in hydrocarboxylation. The remarkable

(15)   (16)

(17a)   (17b)

complex (18) is readily converted into the isomeric 2, 4, 6-triphenyltropone complexes (17a and 17b), of which it is probably the precursor. The liberation of 2, 4, 6-triphenyltropone from these complexes is best achieved by reaction with triphenylphosphine, as thermal decomposition gives exclusively 1, 3, 5-triphenylbenzene. p-Bromophenylacetylene gives an analogous range of products to phenylacetylene[22].

(18)   (19)

It is reported that all three dimethyldiphenylcyclopentadienoneiron tricarbonyl complexes are obtained from 1-phenylpropyne[20]. Similar observations have been made with diphenylbutadiyne[54] and trimethylsilylphenylacetylene. Methyl phenylpropiolate and diiron nonacarbonyl generate the complexes (19) and (20) in addition to the two isomeric derivatives of cyclobutadieneiron tricarbonyl[29].

(20)

(21)

(22)

Acetylene itself also gives a wide range of products with iron carbonyls, the relative amounts depending upon the reaction conditions[35, 55,84,116]. Tropone is obtained as its iron tricarbonyl complex in up to 30% yield. Other products relevant to the present discussion are hydroquinone and complexes (21), (22) and (23). Complex (21) forms a 2:1 adduct with hydroquinone which was originally formulated as Fe $C_{11}H_7O_5$[84]. The complex (22) is also obtained from thiophen and triiron dodecacarbonyl and is converted by carbon monoxide into cyclopentadienone iron tricarbonyl (21). Despite its suggestive structure complex (23) does not serve as a precursor of hydroquinone through replacement of the ring-bonded iron tricarbonyl group by acetylene[20]. Although with acetylene it gives a good yield of hydroquinone, substitution of but-2-yne leads to the formation of tetramethyl rather than dimethylhydroquinone.

(23)

(24)

Irradiation of iron pentacarbonyl in but-2-yne gives the 2, 3, 5, 6-tetramethylbenzoquinoneiron tricarbonyl complex (24) and similar products are obtained with pent-1-yne and hex-1-yne[109]. These complexes are decomposed by air to the quinone and by dilute acid to the hydroquinone. Similarly, irradiation of but-2-yne in a solution containing the isoelectronic manganese pentacarbonyl anion and subsequent acidification gave some 2, 3, 5, 6-tetramethylhydroquinone[64].

Apparently the best source of hydroquinones is the reaction of acetylenes with iron pentacarbonyl in aqueous alkali[84], Table 8. 2. Acry-

213

lic acids are also formed in small amounts. Monosubstituted acetylenes yield 2,5-disubstituted hydroquinones.

(25)

(26)

(27)

Iron carbonyls are not alone in converting acetylenes into cyclopentadienones. Nickel carbonyl converts diphenylacetylene and bis-(p-methoxyphenyl)acetylene into the corresponding cyclopentadienones[16,17]. Cyclopentadienone is probably the precursor of the indan-1-one (25), via its dimer (26), when acetylene is reacted with nickel carbonyl[84]. Diphenylacetylene is converted into tetraphenylcyclopentadienone by tris(acetonitrile)chromium tricarbonyl[112] and by irradiation with π-methylcyclopentadienylmanganese tricarbonyl[111]. Irradiation of acetylenes with π-cyclopentadienylcobalt dicarbonyl produces the complexes (27, R = CH_3, CF_3, C_6H_5)[19, 64]. Tetrasubstituted cyclopentadienone rhodium (I) chloride complexes are formed in the reaction of rhodium dicarbonyl chloride with diphenylacetylene and hex-3-yne[63]. Another product of the latter reaction is tetraethylbenzoquinone. Reaction of bis(trifluoromethyl)acetylene with rhodium dicarbonyl chloride gives free tetrakis(trifluoromethyl)cyclopentadienone[30]. The latter ketone has also been obtained in the form of iron tricarbonyl[19] and π-cyclopentadienylrhodium[30] complexes from reaction of the acetylene with iron pentacarbonyl and π-cyclopentadienylrhodium dicarbonyl respectively.

Terminal acetylenes with a bulky substituent, such as t-butyl or trimethylsilyl, react with bis(tetracarbonylcobalt)mercury forming complexes of structure (28). These are cleaved by halogens giving successively complex (29) and the free cyclopentadienone[59]. In this way monomeric di-t.-butylcyclopentadienone has been prepared. Although the orientation was not established it is probably the 2,5-isomer. Alternatively the acetylenedicobalt hexacarbonyl complex (30) may be prepared first and further reacted with the acetylene giving (31). Again the complex may be degraded with halogen via (29) to the free cyclopentadienone. This type of reaction has been used to prepare 2,3,5-tri-t.-butylcyclopentadienone[46]. The initially

TABLE 8.2 Formation of Hydroquinones[84]

Acetylene	Hydroquinone	Yield, %
Propyne	2, 5-Dimethylhydroquinone	30
But-2-yne	2, 3, 5, 6-Tetramethylhydroquinone	15
Phenylacetylene	2, 5-Diphenylhydroquinone	22
Prop-1-yn-3-ol	2, 5-Di-(hydroxymethyl)-hydroquinone	—
But-2-yn-1, 4-diol dimethyl ether	2, 3, 5, 6-Tetra-(methoxymethyl)-hydroquinone	27.5
Hex-3-yn-2, 5-diol dimethyl ether	2, 3, 5, 6-Tetra-α(α-methoxyethyl)-hydroquinone	2
3-Dimethylaminoprop-1-yne	2, 5-Di-(dimethylaminomethyl)-hydroquinone	4
3-Diethylaminoprop-1-yne	2, 5-Di-(diethylaminomethyl)-hydroquinone	2
Bis-1, 4-dimethylaminobut-2-yne	2, 3, 5, 6-Tetra-(dimethylaminomethyl)-hydroquinone	22

prepared t.-butyl- or di-t.-butyl-acetylenedicobalt hexacarbonyl complex was heated with di-t.-butyl or t.-butyl acetylene respectively. Reaction of the tri-t.-butyl-cyclopentadienone with t.-butylacetylene gave the "overcrowded" 1, 2, 4, 5-tetra-t.-butylbenzene, which was also obtained directly from the preceding reaction.

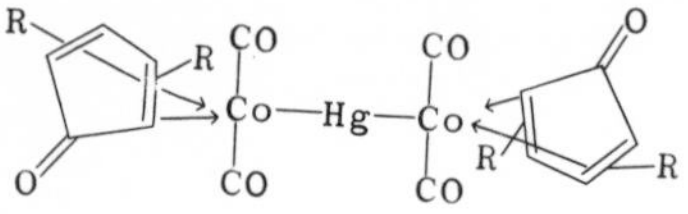

(28)

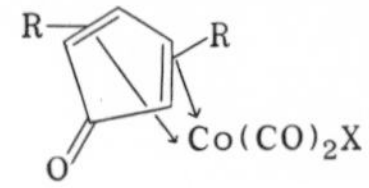

(29)

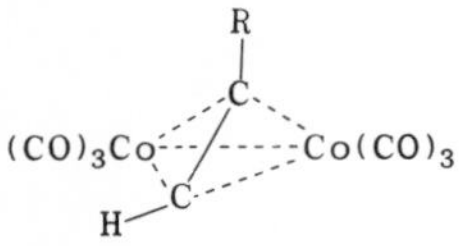

(30)

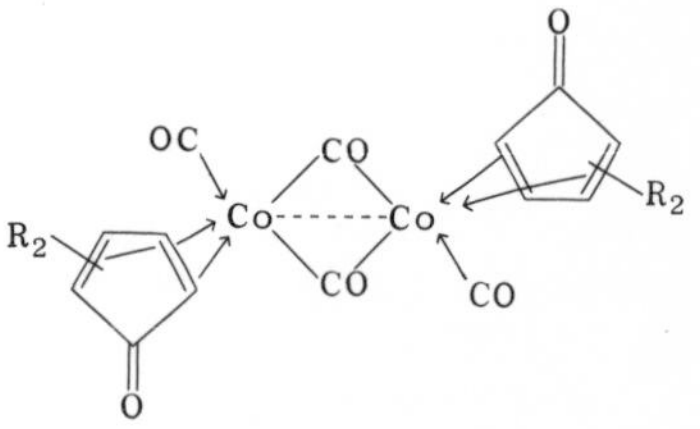

(31)

The reactions depicted in Figure 8.3 cast an interesting light on the foregoing reactions[20,59]. Although the position is somewhat obscured by exchange between the acetylene and the acetylenedicobalt hexacarbonyl complex, the formation of the cyclopentadienone (**32**) from diphenylacetylenedicobalt hexacarbonyl and bis-(trimethylsilyl)acetylene implies the formation of a symmetrical intermediate such as

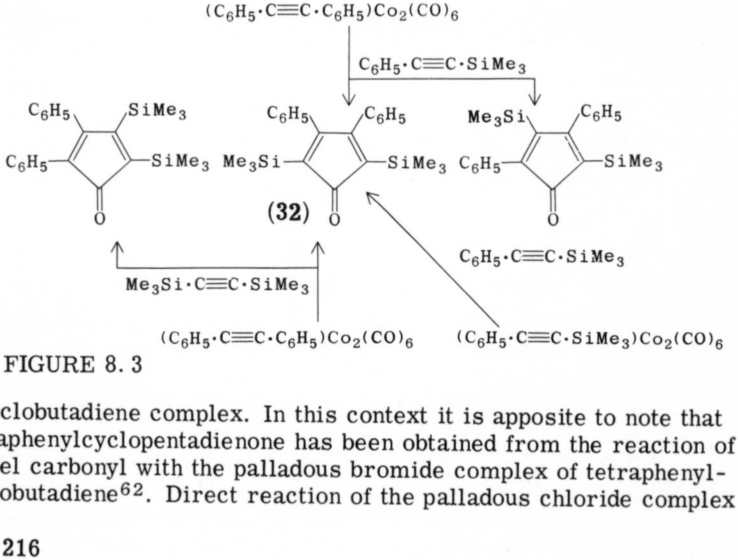

FIGURE 8.3

a cyclobutadiene complex. In this context it is apposite to note that tetraphenylcyclopentadienone has been obtained from the reaction of nickel carbonyl with the palladous bromide complex of tetraphenyl-cyclobutadiene[62]. Direct reaction of the palladous chloride complex of tetraphenylcyclobutadiene with carbon monoxide yields 2, 3, 4, 5-tetraphenylcyclopent-2-en-1-one[114].

A formally analogous transformation can be seen in the reaction of biphenylene with chromium hexacarbonyl[7]. The products obtained are fluorenone and bis-biphenylene-ethylene, which is formed from fluorenone under these conditions.

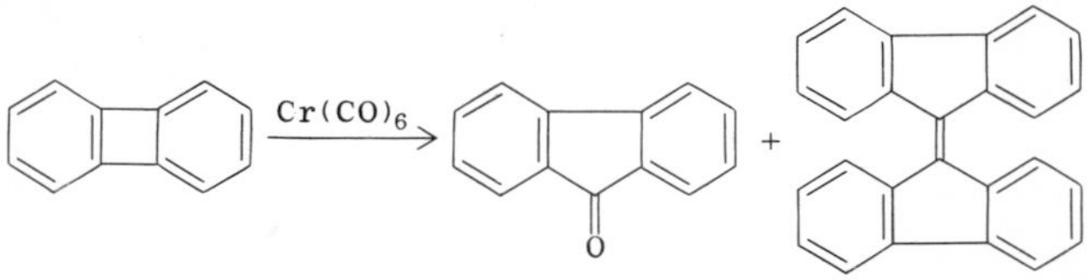

The reaction of acetylenes with dicobalt octacarbonyl in non-hydroxylic solvents takes a different course when carried out at 90-120°C. and 100-1, 000 atmospheres pressure of carbon monoxide, yielding mainly *trans*-bifurandione (33) with smaller amounts of the *cis*-isomer (34)[1,2,3,45,104].

(33)   (34)

The *trans*-form is less stable than the *cis*-form into which it is converted rapidly by acid or base treatment. Substituted acetylenes such as hex-3-yne[104], propyne[1,3,104], hex-1-yne[3,104], as well as phenyl-[3,104], 2-naphthyl-, *p*-chlorophenyl and *o*-methoxyphenylacetylene[104] give substituted bifurandiones. Unsymmetrically substituted acetylenes appear to yield mixtures of positional isomers. In the case of propyne two isomers, (35) and (36), have been isolated in *trans* and *cis* forms[1].

(35)   (36)

The origin of this type of compound is possibly seen in the lactonic complex (37), which is formed from the acetylenedicobalt hexacar-

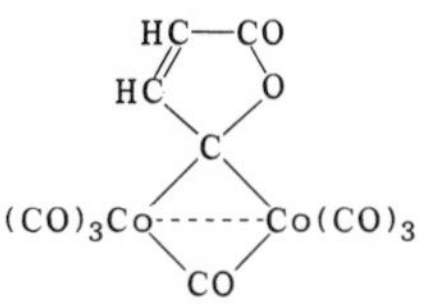

(37)

bonyl complex under similar conditions[66]. Under the same reaction conditions 3-dialkylaminoprop-1-ynes give products formulated as $(38; R = CH_3, C_2H_5)$ and 3-diethylaminobut-1-yne forms (39). [105]

$$R_2N \cdot CO \cdot CH\text{-}CH$$
$$H_2C \qquad C \cdot NR_2$$
$$O$$

(38)

$$(C_2H_5)_2N \cdot CO \cdot CH\text{-}CH$$
$$C_2H_5 \cdot CH \qquad C \cdot N(C_2H_5)_2$$
$$O$$

(39)

Lactonic products are also observed in the reaction of acylcobalt tetracarbonyls with acetylenes[40]. Acetylcobalt tetracarbonyl and hex-3-yne yield the complex (40a), which is hydrogenated to the corresponding γ-butyrolactone derivative. Hydrogenation of the complex analogous to (40a) obtained from acetylene yields γ-valerolactone. Iodine oxidation of the product (40b) from hex-3-yne and benzoylcobalt tetracarbonyl gives a dimer whose structure is probably (41). Carboethoxyacetylcobalt tetracarbonyl with hex-3-yne forms (40c), which on treatment with dicyclohexylethylamine liberates the lactone (42). This reaction can be effected catalytically from ethyl bromoacetate and hex-3-yne in the presence of carbon monoxide and tetracarbonylcobaltate anion. Analogous reactions have been carried out with chloroacetonitrile, and methyl 4-bromo-but-2-enoate which gives (43).

(40) a $R=CH_3$

b $R=C_6H_5$

c $R=CH_2 \cdot CO_2 \cdot C_2H_5$

$$C_2H_5 \cdot C$$
$$C_2H_5 \cdot C \qquad \qquad O$$
$$Co \qquad R$$
$$(CO)_3$$

(41)

(42)

$$C_2H_5 \qquad C_2H_5$$
$$O= \qquad =CH \cdot CO_2C_2H_5$$
$$O$$

(43)

$$C_2H_5 \qquad C_2H_5$$
$$O= \qquad =CH \cdot CH=CH \cdot CO_2C_2H_5$$
$$O$$

A different approach to the cyclopentadienone system is illustrated in Figure 8.4[39]. The reaction of *trans,trans*-hexa-2,4-dienoyl chloride (sorbyl chloride) with sodium tetracarbonylcobaltate at 0°C yields a mixture of the complex (44) and sorbylcobalt tetracarbonyl. The latter is converted by triphenylphosphine into (45), which is cyclised to (46) by heating at 70-80°C. Treatment with triphenylmethyltetrafluoroborate converts (46) into (47). Similarly, *trans*-penta-2,4-dienoyl chloride can be converted into the cyclopentadienone analogue of (47).

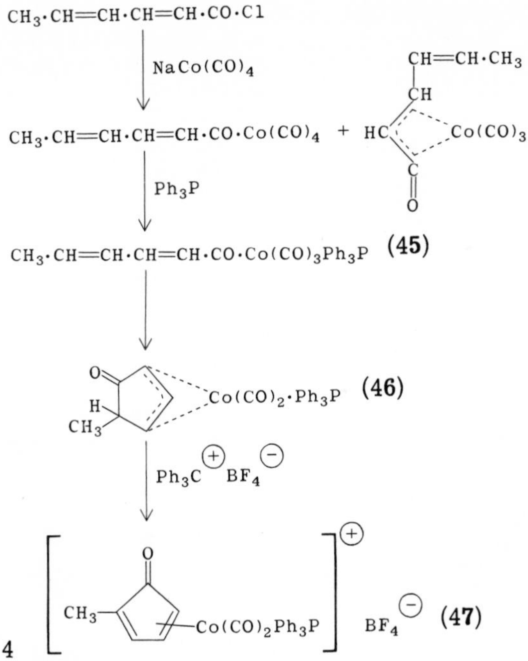

$$CH_3 \cdot CH{=}CH \cdot CH{=}CH \cdot CO \cdot Cl$$

$$CH_3 \cdot CH{=}CH \cdot CH{=}CH \cdot CO \cdot Co(CO)_4 \; + \; HC{\overset{\displaystyle \underset{CH}{\overset{CH{=}CH \cdot CH_3}{}}}{\underset{\displaystyle \underset{O}{C}}{\vphantom{X}}}}Co(CO)_3 \qquad (44)$$

$$CH_3 \cdot CH{=}CH \cdot CH{=}CH \cdot CO \cdot Co(CO)_3 Ph_3 P \quad (45)$$

(46)

(47)

FIGURE 8.4

Tetraphenylallene reacts with dicobalt octacarbonyl at 230-250°C and a carbon monoxide pressure of 150 atmospheres forming the ketones (48) and (49) in addition to 1,1,3-triphenylindene[57].

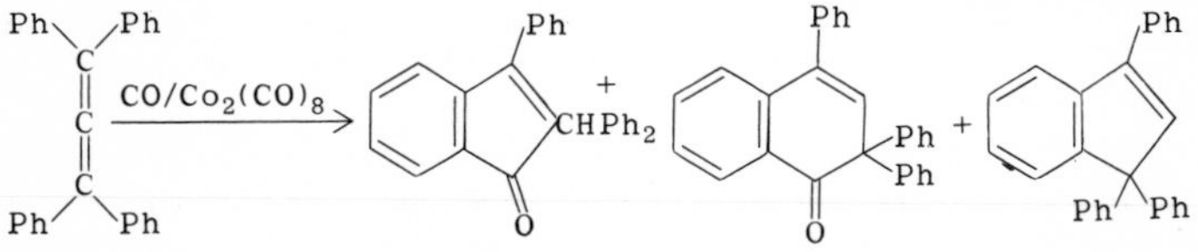

(48), 23%   (49), 17%   42%

1,1,4,4-Tetraphenylbutatriene gives mostly (70%) the indenone (50) with only traces of (51).

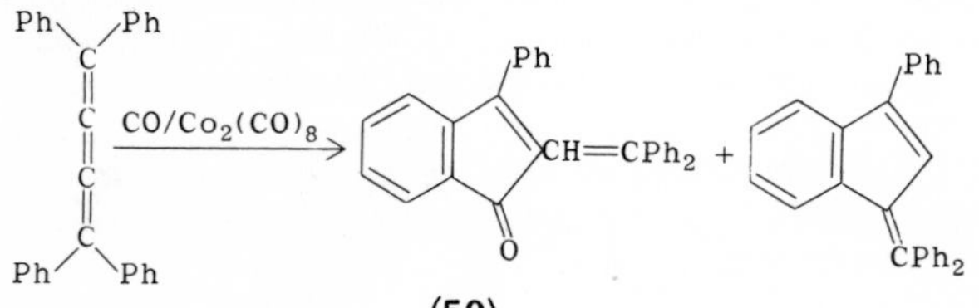

(50)   (51)

219

A logical extension of carbonylation reactions of this type is to compounds containing nitrogen-nitrogen and carbon-nitrogen double bonds. Azobenzene reacts with dicobalt octacarbonyl and carbon monoxide at 170-190°C and 150 atmospheres forming 2-phenylindazolone (52) and small amounts of 3-phenyl-2,4-dioxo-1,2,3,4-tetrahydroquinazoline (53), (54) and diphenylurea[47,48,49,51,69,72]. At 220-230°C the quinazoline (53) becomes the major product with small amounts of diphenylurea and (54). (The original investigators formulated this latter product as (55) but subsequent work shows that (54)

(52)

(53)

(54)

(55)

is the correct structure[67].) Under the same conditions (52) is converted to (53) in 82% yield but similar reactions do not occur with indazole, indazolone, 2-phenylbenzoxazole or 2-phenylbenzimidazole[51,70]. Polar solvents such as ethanol, tetrahydrofuran and water inhibit the reaction[51,72]. Benzene is one of the best solvents. Solvents such as hexane, cyclohexane and tetralin which can act as hydrogen donors cause the formation of larger amounts of diphenylurea[49,51,72]. In all of the cases reported (Table 8.3) cyclisation always occurs onto the substituted ring. Iron pentacarbonyl is reported to give lower yields[47,51,69,72] and the reaction of azobenzene with nickel carbonyl in cyclohexane at 250°C yields mainly (54) together with aniline, diphenylurea and (53).[51,69,80]

The conversion of (52) to (53) is of some interest. It has been reported that carbonyl exchange occurs between ^{14}CO and diphenylurea at 230°C in the presence of dicobalt octacarbonyl[117]. This suggests that the diarylureas obtained as by-products in the carbonylation of azobenzenes may arise *via* reduction to the diarylhydrazine and carbonylation. The conversion of hydrazobenzene to diphenylurea has been effected under the reaction conditions[49,51,72]. Hydrazine itself reacts with carbon monoxide in the presence of iron pentacarbonyl at 45°C and 900 atmospheres producing semicarbazide, while at 100°C and 500 atmospheres urea is also formed[103].

TABLE 8.3 Carbonylation of Azobenzenes $R-N=N-R^1$

R	R^1	Product	Yield, %	References
C_6H_5	C_6H_5	2-Phenylindazolone	50	47, 48, 51, 69, 72
$p\text{-}CH_3C_6H_4$	C_6H_5	5-Methyl-2-phenylindazolone	35	48, 51, 72
$p\text{-}Cl.C_6H_4$	C_6H_5	5-Chloro-2-phenylindazolone	24	48, 51, 69, 72
$p\text{-}(CH_3)_2N.C_6H_4$	C_6H_5	5-Dimethylamino-2-phenylindazolone	80	48, 51, 69, 72
C_6H_5	C_6H_5	3-Phenyl-2, 4-dioxo-1, 2, 3, 4-tetrahydroquinazoline	80	47, 48, 51, 69, 70
$p\text{-}CH_3.C_6H_4$	C_6H_5	6-Methyl-3-phenyl-2, 4-dioxo-1, 2, 3, 4-tetrahydro-quinazoline	36	72
$o\text{-}CH_3.C_6H_4$	C_6H_5	8-Methyl-3-phenyl-2, 4-dioxo-1, 2, 3, 4-tetrahydro-quinazoline	26	
$p\text{-}Cl.C_6H_4$	C_6H_5	6-Chloro-3-phenyl-2, 4-dioxo-1, 2, 3, 4-tetrahydro-quinazoline	45	48, 51, 69, 72
$p\text{-}(CH_3)_2NC_6H_4$	C_6H_5	6-Dimethylamino-3-phenyl-2, 4-dioxo-1, 2, 3, 4-tetra-hydroquinazoline	18	48, 51, 69, 72
$p\text{-}CH_3C_6H_4$	$p\text{-}CH_3C_6H_4$	6-Methyl-3-(p-tolyl)-2, 4-dioxo-1, 2, 3, 4-tetrahydro-quinazoline	40	48, 51, 72
$p\text{-}Cl.C_6H_4$	$p\text{-}Cl.C_6H_4$	6-Chloro-3-(p-chlorophenyl)-2, 4-dioxo-1, 2, 3, 4-tetrahydroquinazoline	17	48, 51, 72
$p\text{-}CH_3O.C_6H_4$	$p\text{-}CH_3O.C_6H_4$	6-Methoxy-3-(p-methoxyphenyl)-2, 4-dioxo-1, 2, 3, 4-tetrahydroquinazoline	28	48, 51, 72
$p\text{-}NC.C_6H_4$	C_6H_5	No appreciable reaction	—	48, 51, 72
α-Naphthyl	α-Naphthyl	No appreciable reaction	—	48, 51, 72
β-Naphthyl	β-Naphthyl	No appreciable reaction	—	48, 51, 72

221

TABLE 8.4 Carbonylation of Amines

Amine	Catalyst	Products	References
Ammonia	$Re(CO)_5Cl$	Formamide	44
	$Ru(CO)_2I_2$	Formamide	42
Methylamine	$Hg[Co(CO)_4]_2$	N-Methylformamide	44
iso-Propylamine	$Ni(CO)_4$	N-iso-Propylformamide	4
Butylamine	$Ni(CO)_4$	N-Butylformamide and N, N'-dibutylurea	4, 82, 83
	$CH_3Mn(CO)_5$	N, N'-Dibutylurea (33%) and N-butylformamide	25
	$Mn_2(CO)_{10}$	N, N'-Dibutylurea (52%) and N-butylformamide	25
Amylamine	$Ni(CO)_4$	N-Amylformamide	4
n-Hexylamine	$Ni(CO)_4$	N-n-Hexylformamide	4
Cyclohexylamine	$CH_3Mn(CO)_5$	N, N'-Dicyclohexylurea (25%) and N-cyclohexyl-formamide	25
	$Mn_2(CO)_{10}$	N, N'-Dicyclohexylurea (58%) and N-cyclohexyl-formamide	25
Aniline	$Ni(CO)_4$	N, N'-Diphenylurea and formanilide	4, 82
	$Co_2(CO)_8$	N, N'-Diphenylurea	83
	$Mn_2(CO)_{10}$	N, N'-Diphenylurea (6%)	25
Dimethylamine	$Co_2(CO)_8$	Dimethylformamide and tetramethylurea	110
Diethylamine	$Ni(CO)_4$	Diethylformamide	4
Dibutylamine	$Ni(CO)_4$	Dibutylformamide	82, 83

Amine	Reagent	Product	References
Di-isobutylamine	$Ni(CO)_4$	Di-iso-butylformamide	4, 83
Diphenylamine	$Ni(CO)_4$	Diphenylformamide and tetraphenylurea	4
Pyrrolidine	$Ni(CO)_4$	N-Formylpyrrolidine	82, 83
Piperidine	$Fe(CO)_5$	N-Formylpiperidine	43
	$Co_2(CO)_8$	N-Formylpiperidine	110
	$Ni(CO)_4$	N-Formylpiperidine	4, 5, 43, 82, 83
2-Methylpiperidine	$Ni(CO)_4$	N-Formyl-2-methylpiperidine	6
3-Methylpiperidine	$Ni(CO)_4$	N-Formyl-3-methylpiperidine	6
Anabasine	$Ni(CO)_4$	N-Formylanabasine	6
Morpholine	$Fe(CO)_5$	N-Formylmorpholine	43
	$Ni(CO)_4$	N-Formylmorpholine	43, 82, 83
Perhydrocarbazole	$Ni(CO)_4$	N-Formylperhydrocarbazole	82, 83
Hexamethyleneimine	$Co_2(CO)_8$	N-Formylhexamethyleneimine	83
Trimethylamine	$Co_2(CO)_8$	Dimethylformamide	83
Tributylamine	$Co_2(CO)_8$	Dibutylformamide	83
	$Ni(CO)_4$	Dibutylformamide	82
N,N-Diethylaniline	$Ni(CO)_4$	N-Ethyl-N-phenylformamide	4
	$Ni(CO)_4$	N-Ethyl-N-phenylpropionamide	82
N,N-Dimethyl-β-naphthylamine	$Ni(CO)_4$	N-Methyl-N-β-naphthylacetamide	82

Alternatively, the diarylureas may arise from carbonylation of the arylamine formed by hydrogenolysis. Whereas aniline reacts with metal carbonyls forming diphenylurea and lesser amounts of formanilide, aliphatic amines yield mostly the N-formyl derivatives (Table 8. 4). However, 1, 3-dialkylureas become the major products in the reaction of primary aliphatic amines with manganese carbonyls[25]. Aliphatic tertiary amines react with the loss of an alkyl group forming N, N-dialkylformamides[82,83], but tertiary aromatic amines such as N, N-dimethyl-β-naphthylamine and N, N-diethylaniline react as shown.

Carbonylation of amines at 65-85°C and 1. 4 atmospheres using an equimolar amount of palladous chloride gives good yields of isocyanates, $R = C_4H_9$, 49%; $R = C_6H_5$, 68%[108]

$$R.NH_2 + CO + Pd Cl_2 \longrightarrow R.NCO + Pd + 2HCl$$

Under more vigorous conditions, 180°C and 100 atmospheres, n-decylamine yields 1, 3-didecylurea and N, N'-didecyloxamide[113].

The carboxylation of compounds containing a carbon-nitrogen double bond has been extensively investigated. Schiff bases are converted into phthalimidines by reaction with carbon monoxide and dicobalt octacarbonyl in benzene at 200-230°C and 100 to 200 atmospheres (Table 8. 5). N-Benzylideneaniline for example gives an 84% yield of 2-phenylphthalimidine (56). Iron pentacarbonyl also catalyses the reaction but less effectively[72,74]. Di-iron nonacarbonyl reacts with Schiff bases under mild conditions forming complexes of type (57). These are oxidised to phthalimidines by ferric chloride[8]. Nickel carbonyl is reported to be catalytically inactive[68,72,74], although its use has been claimed in a patent[81]. In this reaction also polar solvents have an inhibiting effect[72]. The Schiff bases are reduced to secondary amines in the presence of hydrogen[52,72,73].

(56)   (57)

The factors controlling the direction of cyclisation are not obvious; only a few Schiff bases (58) derived from m-substituted benzaldehydes have been employed. When X is dimethylamino or hydroxyl the phthalimidine (59) is obtained in good yield, but when X is meth-

TABLE 8.5 Carbonylation of Schiff Bases to Phthalimidines

R	R¹	R²	Phthalimidine	Yield, %	References
Phenyl	Phenyl	H	2-Phenyl	80	52, 68, 72, 74, 76, 81
Phenyl	o-Tolyl	H	2-o-Tolyl	53	52
Phenyl	p-Tolyl	H	2-p-Tolyl	88	52
Phenyl	2,6-Dimethylphenyl	H	2-(2,6-Dimethylphenyl)	50	50, 52
Phenyl	2,6-Diethylphenyl	H	2-(2,6-Diethylphenyl)	36	50, 52
Phenyl	p-Methoxyphenyl	H	2-p-Methoxyphenyl	86	72, 75, 76
Phenyl	p-Hydroxyphenyl	H	2-p-Hydroxyphenyl	65	72, 75, 76
Phenyl	p-Chlorophenyl	H	2-p-Chlorophenyl	75	72, 75, 76
Phenyl	p-Diethylaminophenyl	H	2-p-Diethylaminophenyl	—	81
Phenyl	1-Naphthyl	H	2-(1-Naphthyl)	—	81
Phenyl	Methyl	H	2-Methyl	49	72, 75, 76, 81
Phenyl	Cyclohexyl	H	2-Cyclohexyl	—	50
Phenyl	Benzyl	H	2-Benzyl	82	72, 75, 76
Phenyl	1,2-Ethylene	H	1,2-Bis-(2-phthalimidinyl)ethylene	—	81
p-Dimethylaminophenyl	Phenyl	H	2-Phenyl-6-dimethylamino	82	75, 76, 81
p-Hydroxyphenyl	Phenyl	H	2-Phenyl-6-hydroxy	77	68, 75, 76
p-Methoxyphenyl	Phenyl	H	2-Phenyl-6-methoxy	—	81

R - CR² = N - R¹

(Continued overleaf)

TABLE 8.5 (continued)

$R - CR^2 = N - R^1$

R	R^1	R^2	Phthalimidine	Yield, %	References
p-Chlorophenyl	Phenyl	H	2-Phenyl-6-chloro	—	81
o-Methoxyphenyl	Phenyl	H	2-Phenyl-4-methoxy	18	72, 75, 76
m-Methoxyphenyl	Phenyl	H	2-Phenyl-5-methoxy	5	72, 75, 76
m-Hydroxyphenyl	Phenyl	H	2-Phenyl-7-hydroxy	77	72
m-Dimethylamino	Phenyl	H	2-Phenyl-7-dimethylamino	82	72
1-Naphthyl	Phenyl	H	2-Phenyl-4, 5-benzo	96	68, 72, 75, 76
2-Naphthyl	Phenyl	H	2-Phenyl-5, 6-benzo	80	68, 72, 75, 76
Phenyl	Phenyl	Methyl	2-Phenyl-3-methyl	89	50, 52, 72, 75, 76, 77, 81
Phenyl	o-Tolyl	Methyl	2-o-Tolyl-3-methyl	64	50, 52
Phenyl	Phenyl	Phenyl	2, 3-Diphenyl	97	50, 52, 72, 75, 76, 77, 81

oxyl only a low yield of (60) is produced. It seems possible that steric effects are important as 2-naphthaldehyde anil gives a high yield of (61) by cyclising at position 3, whereas electronic factors would favour cyclisation at position 1. Steric hindrance by ortho substituents such as 2-methyl, 2, 6-dimethyl or 2, 6-diethyl on $R' = C_6H_5$ lowers the rate of carbonylation of $RCR^2 = N.R'$ with $R = H$, although a p-methyl group enhances it[72]. When $R = R' = C_6H_5$ the

CH$_2$ ... image ...

(59)   (58)   (60)

(61)

nature of $R^2 = H$, CH_3 or C_6H_5 had no appreciable effect on the rate. No steric effect from the *ortho* substituents is observed for the reduction of the Schiff bases.

The conversion of benzonitrile into 2-benzylphthalimidine[89,90] can be viewed as a special case of the carbonylation of a Schiff base. As far higher yields (41%) are obtained in the presence of hydrogen and the addition of benzylamine it seems that prior reduction of the benzonitrile occurs.

$C_6H_5 \cdot C \equiv N$ — [$C_6H_5 \cdot CH = NH$ / $C_6H_5 \cdot CH_2 \cdot NH_2$] → $C_6H_5 \cdot CH = N \cdot CH_2 C_6H_5$

(62)

The addition of pyridine also increases the yield of (62). The major by-products particularly under less favourable conditions are benzamide and dibenzylurea. As anticipated substitution of deuterium for hydrogen in the carbonylation of benzonitrile leads to equal incorporation of deuterium into both methylene groups. This excludes a previously proposed mechanism[52,96] in which cyclisation is accompanied by intramolecular transfer of the displaced hydrogen to the double bond.

Oximes are less satisfactory as phthalimidine precursors and a variety of other products have been obtained. Benzaldoxime is variously reported as giving benzyl and dibenzylurea[94] or benzamide[72, 76]. Acetophenone oxime produces 3-methylphthalimidine[88], and the oxime of 2-acetylnaphthalene yields ureas together with a large amount of (63) and a small quantity (10%) of (64)[91]. It appears that N-hydroxy or N-methoxy phthalimidines readily undergo hydrogenolysis. The benzoquinoline (63) is also formed in the absence of dicobalt octacarbonyl and carbon monoxide[87]. Phenyl benzyl ketoxime produces 3-benzylphthalimidine and small amounts of the isocarbostyril (65).[88] However, the oximes of dibenzylketone and 1-phenyl-butan-3-one do not undergo similar reactions, but are reduced to the amines and converted to formyl derivatives and ureas[97]. Both benzophenone oxime and its O-methyl ether yield 3-phenylphthalimidine in up to 80% yield[86,88,97]. The same product is also obtained in similar yield from the oxime of o-benzoylbenzoic acid[88] and suggests that cyclisation and decarboxylation are concerted. Subjection of α-benzildioxime N,N'-dimethyl ether to the carbonylation conditions produced tetraphenylpyrazine[86].

(63)   (64)   (65)

Carbonylation of suitable semicarbazones also leads to formation of phthalimidine derivatives. Benzaldehyde semicarbazone is converted under the reaction conditions, 240°C and ca. 290 atmospheres, into the azine which in turn gives (66), (67) and (68) together with a compound provisionally identified as the benzodiazepine (69).[93]

$C_6H_5 \cdot CH=N \cdot N=CH \cdot C_6H_5 \longrightarrow$

(66) R=H, 13%   (69), 12%

(67) R=CH$_2$C$_6$H$_5$, 16%

(68) R=CO·NH·CH$_2$C$_6$H$_5$, 5%

The products from benzophenone semicarbazone depend on the reaction temperature and it is only at higher temperatures (ca. 240°C) that phthalimidine formation occurs[92]. At lower temperatures reduction and azine formation predominate[92].

$(C_6H_5)_2C=N \cdot NH \cdot CONH_2 \longrightarrow$ ... $+ (C_6H_5)_2CH \cdot NH \cdot CO \cdot NH_2$
$+ ((C_6H_5)_2CHNH)_2CO$

(**70**) R=H

(**71**) R=CO·NH·CH(C$_6$H$_5$)$_2$

The azine gives (**70**) under these conditions and $(C_6H_5)_2C = N.NH.CO.NH CH(C_6H_5)_2$, which is a product at lower temperatures also forms (**70**) and (**71**) at 240°C.

The conversion of phenylhydrazones to phthalimidines has been extensively investigated. Benzaldehyde phenylhydrazone yields 2-phenylphthalimidine (50%) accompanied by benzonitrile, benzaldehyde and diphenylurea[96]. The origin of the phthalimidine nitrogen has been confirmed by use of nitrogen-15 labelled compounds.

The intramolecular nature of the reaction is shown by the absence of crossed products when the phenylhydrazones (**72**) and (**73**) are carbonylated together.

(**72**)

(**73**)

The clear differentiation between the two nitrogen atoms is not observed in the carbonylation of phenylhydrazones of arylketones. Acetophenone phenylhydrazone gives two products (**74**) and (**75**) resulting from loss of either nitrogen[96], while benzophenone gives only (**76**) in 70% yield. However, at lower temperatures the principal product is 3-phenylphthalimidine, indicating that the addition of the phenylcarbamyl group occurs subsequent to phthalimidine formation

229

(74)   (75)

and apparently excluding formation of (76) by carbon monoxide insertion into the nitrogen-nitrogen bond of a 2-phenylaminophthalimidine
(77) precursor.

(76)   (77)

Whereas in the carbonylation of azobenzenes cyclisation always
occurs onto the substituted ring, carbonylation of 4-methylbenzophe-
none phenylhydrazone yields both isomeric phthalimidines[95].

20%   20%

The carbonylation of benzyl phenyl ketone phenylhydrazone to 3-ben-
zyl analogues of (74) and (75) is accompanied by the formation of 2,
3-diphenylindole[98]. The latter is formed in 80% yield in the absence
of carbon monoxide and dicobalt octacarbonyl. In the case of diben-
zyl ketone phenylhydrazone formation of 2-benzyl-3-phenylindole
occurs to the exclusion of phthalimidine formation.

Reference was made earlier in this chapter to the formation of
N, N'-didecyloxamide by carbonylation of n-decylamine catalysed by
palladous chloride. Reactions such as this, which entail coupling of
two acyl groups, have been observed in other carbonylation reactions.
The first example of this type of reaction was encountered in the
carbonylation of $trans$-[$Me_2Pt(Et_3P)_2$] which produced diacetyl[18].
A reaction of synthetic importance is the formation of arils from
the reaction of aryl iodides with nickel carbonyl in tetrahydrofuran
at 40°C.[12] Two hindered aryl iodides yield enediol esters (78) in
addition to the arils (Table 8.6).

TABLE 8.6 Formation of Arils[12]

Aryl Iodide	Aril, %	Enediol Diester, %
Phenyl	80	0
p-Chlorophenyl	47	0
m-Tolyl	50	0
o-Tolyl	10	26
1-Naphthyl	35	29
Mesityl	0	0

Under the same conditions benzoyl chloride and bromide are converted into 1,2-dibenzoyloxystilbene (**78**, Ar $= C_6H_5$). The enediol esters result from acylation of the dianion formed by reaction of the aril with nickel carbonyl. In the case of reactions in tetrahydrofuran only 45% of the enediol diester is obtained as the cis-isomer but in hexane over 90% of the cis-form is generated[11,12]. The formation of the enediol diesters in the case of the hindred aryl iodides is difficult to rationalise. It seems likely that the acylating species is Ar.CO Ni(CO)$_2$I. The observation[26] that complexes of the type [Ar Ni(Et$_3$P)$_2$Br] are only stable enough for isolation when the aryl group is sterically hindered, as in the case of 1-naphthyl or o-tolyl, is undoubtedly relevant.

$$\text{Ar I} \xrightarrow{\text{Ni(CO)}_4} \text{Ar·CO·CO·Ar} \longrightarrow \underset{\text{Ar} \quad \text{Ar}}{\overset{\text{ArCO·O} \quad \text{O·CO·Ar}}{\text{C}=\text{C}}}$$

(78)

Closely related intermediates are involved in the formation of arils and aroins by the reaction of organolithium compounds with nickel carbonyl[78,99] and iron pentacarbonyl[101,102]. Direct reaction of organolithium compounds with carbon monoxide produces ketones[102]. The assumption that the reactions with metal carbonyls involve lithium aroylmetal carbonylates is supported by the isolation of the benzoyltungsten carbonylate anion, $[C_6H_5.CO.W(CO)_5]^-$, as its tetramethylammonium salt from the reaction of phenyllithium with tungsten hexacarbonyl[33,65]. Protonation of this complex yields benzaldehyde.

The reaction of p-tolyllithium with nickel carbonyl in equimolar amounts at $-70°C$ gives a salt-like air sensitive black powder which is apparently Li$^+$[p.CH$_3$C$_6$H$_4$CO Ni(Co)$_3$]$^-$[99]. Hydrolysis of this compound yields p-toluoin and thermal decomposition generates p-

TABLE 8.7 Formation of Acyloins

Organometallic Compound	Metal Carbonyl	Products	Yield, %	References
n-Propyl magnesium bromide	Ni(CO)$_4$	n-Butyroin	50	13
iso-Propyl magnesium bromide	Ni(CO)$_4$	iso-Butyroin	35	13
n-Butyllithium	Ni(CO)$_4$	Dibutyl ketone	29	78
n-Butylmagnesium bromide	Ni(CO)$_4$	n-Valeroin	50	13
Phenyllithium	Ni(CO)$_4$	Benzoin	36	78
	Fe(CO)$_5$	Benzoin	25	102
Phenyl magnesium bromide	Ni(CO)$_4$	Benzoin	70	13, 99
p-Tolyllithium	Ni(CO)$_4$	p-Toluoin (71%) and p-tolil (20%)	—	78, 99
m-Tolyllithium	Ni(CO)$_4$	m-Toluoin	61	78
o-Tolyllithium	Ni(CO)$_4$	o-Toluoin	63	99
		o-Tolil	42	99
p-Anisyllithium	Ni(CO)$_4$	p-Anisoin (15%) and p-anisil (23%)	—	78
2,6-Dimethylphenyl lithium	Ni(CO)$_4$	2, 2', 6, 6'-Tetramethylbenzil (23%) and 2, 6-dimethylbenzaldehyde (22%)	—	78

tolil. Addition of an equimolar amount of bromine to the complex followed by ethanolysis forms p-tolil (72.5%) and ethyl p-toluate (12%). The intermediate (79) here is obviously identical with that postulated for the reaction of aryl iodides with nickel carbonyl.

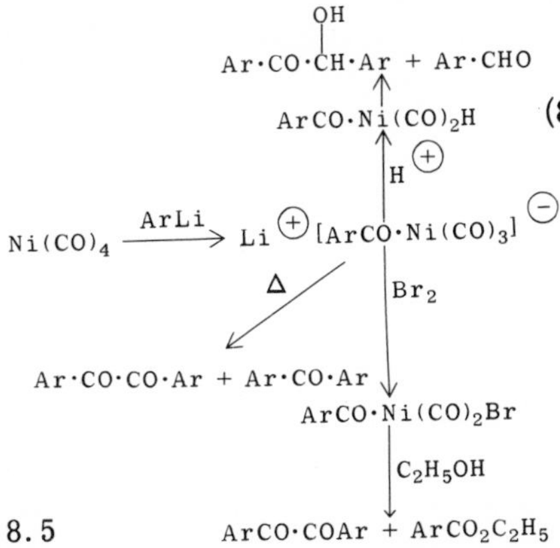

FIGURE 8.5

Reaction of lithium p-toluoylnickel carbonylate with benzoyl chloride gave α, α'-dibenzoyloxy-4, 4'-dimethylstilbene, and with benzyl chloride α-benzyl-p-toluoin (73%). These reactions can be rationalised as in Figure 8.5 (Ar = p-tolyl). The use of a higher ratio of organolithium to nickel carbonyl results in the formation of other products such as diarylketones rather than aroins. Good yields of aroins have also been obtained from the reaction of Grignard reagents with nickel carbonyl[13,99]. The reaction of aryl-lithiums with iron pentacarbonyl gives mainly aldehydes[100,101].

REFERENCES

1. Albanesi, G., Chimica e Industria, 1964, 46, 1169.

2. Albanesi, G. and Tovaglieri, M., Chimica e Industria, 1959, 41, 189.

3. Albanesi, G. and Tovaglieri, M., Swiss P. 376, 900; Chem. Abs., 1964, 61, 14538.

4. Aliev, Y. Y. and Romanova, I. B., Neftekhimiya, Akad. Nauk Turkm. S.S.R., 1963, 204; Chem. Abs., 1964, 61, 6913.

5. Aliev, Y. Y., Romanova, I. B. and Freidlin, L. K., Uzbek. khim. Zhur., 1960, 72; Chem. Abs., 1961, 55, 10444.

6. Aliev, Y. Y., Romanova, I. B. and Freidlin, L. K., Uzbek. khim. Zhur., 1962, 6, 58; Chem. Abs., 1963, 59, 3882.

7. Atkinson, E. R., Levins, P. L. and Dickelman, T. E., Chem. and Ind., 1964, 934.

8. Bagga, M. M., Pauson, P. L., Preston, F. J. and Reed, R. I., Chem. Comm., 1965, 543.

9. Bannister, W. D., Green, M. and Haszeldine, R. N., Proc. Chem. Soc., 1964, 370.

10. Bannister, W. D., Green, M. and Haszeldine, R. N., J. Chem. Soc., 1966, 194.

11. Bauld, N. L., J. Amer. Chem. Soc., 1962, **84**, 4345.

12. Bauld, N. L., Tetrahedron Letters, 1963, 1841.

13. Benton, F. L., Voss, M. C. and McKusker, P. A., J. Amer. Chem. Soc., 1945, **67**, 82.

14. Bertrand, J. A., Aldridge, C. L., Husebye, S. and Jonassen, H. B., J. Org. Chem., 1964, **29**, 790.

15. Bird, C. W., Cookson, R. C. and Hudec, J., Chem. and Ind. 1960, 20.

16. Bird, C. W. and Hollins, E. M., unpublished work.

17. Bird, C. W. and Hudec, J., Chem. and Ind., 1959, 570.

18. Booth, G. and Chatt, J., Proc. Chem. Soc., 1961, 67.

19. Boston, J. L., Sharp, D. W. A. and Wilkinson, G., J. Chem. Soc., 1962, 3488.

20. Braye, E. H., Hoogzand, C., Hübel, W., Krüerke, U., Merényi, R. and Weiss, E., "Advances in the Chemistry of Coordination Compounds", Proc. 6th. Intern. Conf. Coordn. Chem. Detroit, 1961, 190.

21. Braye, E. H. and Hübel, W., Chem. and Ind., 1959, 1250.

22. Braye, E. H. and Hübel, W., J. Organometallic Chem., 1965, **3** 25.

23. Braye, E. H. and Hübel, W., J. Organometallic Chem., 1965, **3**, 38.

24. Braye, E. H., Hübel, W., and Caplier, I., J. Amer. Chem. Soc., 1961, **83**, 4406.

25. Calderazzo, F., Inorg. Chem., 1965, **4**, 293.

26. Chatt, J. and Shaw, B. L., J. Chem. Soc., 1960, 1718.

27. Colleuille, Y. and Perras, P., Fr. P. 1, 352, 841; Chem. Abs., 1964, **61**, 593.

28. Cookson, R. C., Henstock, J. and Hudec, J., J. Amer. Chem. Soc., 1966, **88**, 1059.

29. Dahl, L. F., Doedens, R. J., Hübel, W. and Nielsen, J., J. Amer. Chem. Soc., 1966, **88**, 446.

30. Dickson, R. S. and Wilkinson, G., J. Chem. Soc., 1964, 2699.

31. Ecke, G. G., U.S.P. 3, 149, 135; Chem. Abs., 1965, **62**, 4054.

32. Eischens, R. P., Francis, S. A. and Pliskin, W. A., J. Phys. Chem., 1956, **60**, 194.

33. Fischer, E. O. and Maasböl, A., Angew. Chem. Internat. Edn., 1964, **3**, 580.

33a. Green, M. and Hancock, R. I., Chem. Comm., 1966, 572.

34. Green, M. and Lucken, E. A. C., Helv. Chim. Acta, 1962, **45**, 1870.

35. Green, M. L. H., Pratt, L. and Wilkinson, G., J. Chem. Soc., 1960, 989.

36. Heck, R. F., J. Amer. Chem. Soc., 1963, **85**, 3116.

37. Heck, R. F., J. Amer. Chem. Soc., 1963, **85**, 3381.

38. Heck, R. F., J. Amer. Chem. Soc., 1963, **85**, 3383.

39. Heck, R. F., J. Amer. Chem. Soc., 1963, **85**, 3387.

40. Heck, R. F., J. Amer. Chem. Soc., 1964, **86**, 2819.

41. Heck, R. F., U.S.P. 3, 137, 715; Chem. Abs., 1964, **61**, 8194.

42. Hieber, W. and Heusinger, H., J. Inorg. Nuclear Chem., 1957, **4**, 179.

43. Hieber, W. and Kahlen, N., Chem. Ber., 1958, **91**, 2223.

44. Hieber, W. and Schuster, L., Z. anorg. Chem., 1956, **287**, 214.

45. Holmquist, H. E., U.S.P. 2, 835, 710; Chem. Abs., 1959, **53**, 3064.

46. Hoogzand, C. and Hübel, W., Tetrahedron Letters, 1961, 637.

47. Horiie, S., J. Chem. Soc. Japan., 1958, **79**, 499.

48. Horiie, S., J. Chem. Soc. Japan., 1959, **80**, 1038.

49. Horiie, S., J. Chem. Soc. Japan., 1959, **80**, 1040.

50. Horiie, S., J. Chem. Soc. Japan., 1959, **80**, 1043.

51. Horiie, S. and Murahashi, S., Bull. Chem. Soc. Japan, 1960, **33**, 88.

52. Horiie, S. and Murahashi, S., Bull. Chem. Soc. Japan, 1960, **33**, 247.

53. Hübel, W. and Braye, E. H., J. Inorg. Nuclear Chem., 1959, **10**, 250.

54. Hübel, W. and Merényi, R., Chem. Ber., 1963, **96**, 930.

55. Hübel, W. and Weiss, E., Chem. and Ind., 1959, 703.

56. Jones, E. R. H., Wailes, P. C. and Whiting, M. C., J. Chem. Soc., 1955, 4021.

57. Kim. P. and Hagihara, N., Bull. Chem. Soc. Japan, 1965, **38**, 2022.

58. Klemchuk, P. P., U.S.P. 2, 995, 607; Chem. Abs., 1962, **56**, 1363.

59. Krüerke, U. and Hübel, W., Chem. Ber., 1961, **94**, 2829.

60. Lansbury, P. T. and Meschke, R. W., J. Org. Chem., 1959, **24**, 104.

61. Leto, J. R. and Cotton, F. A., J. Amer. Chem. Soc., 1959, **81**, 2970.

62. Maitlis, P. M. and Games, M. L., J. Amer. Chem. Soc., 1963, **85**, 1887.

63. Maitlis, P. M. and McVey, S., J. Organometallic Chem., 1965, **4**, 254.

64. Markby, R., Sternberg, H. W. and Wender, I., Chem. and Ind., 1959, 1381.

65. Mills, O. S. and Redhouse, A. D., Angew. Chem. Internat. Edn., 1965, **4**, 1082.

66. Mills, O. S. and Robinson, G., Proc. Chem. Soc., 1959, 156.

67. Mosby, W. L., Chem. and Ind., 1957, 17.

68. Murahashi, S., J. Amer. Chem. Soc., 1955, **77**, 6403.

69. Murahashi, S. and Horiie, S., J. Amer. Chem. Soc., 1956, **78**, 4816.

70. Murahashi, S. and Horiie, S., Jap. P. 2319('59); Chem. Abs., 1960, **54**, 11061.

71. Murahashi, S. and Horiie, S., Jap. P. 3366('59); Chem. Abs., 1960, **54**, 14193.

72. Murahashi, S. and Horiie, S., Ann. Report Sci. Works, Fac. Sci., Osaka Univ., 1959, **7**, 89.

73. Murahashi, S. and Horiie, S., Bull. Chem. Soc. Japan, 1960, **33**, 78.

74. Murahashi, S., Horiie, S. and Jo, T., J. Chem. Soc. Japan, 1958, **79**, 72.

75. Murahashi, S., Horiie, S. and Jo, T., J. Chem. Soc. Japan, 1958, **79**, 75.

76. Murahashi, S., Horiie, S. and Jo, T., Bull. Chem. Soc. Japan, 1960, **33**, 81.

77. Murahashi, S., Horiie, S. and Shiro, T., Jap. P. 9586('59); Chem. Abs., 1960, **54**, 7655.

78. Myeong, S. K., Sawa, Y., Ryang, M. and Tsutsumi, S., Bull. Chem. Soc. Japan, 1965, **38**, 330.

79. Pettit, R., J. Amer. Chem. Soc., 1959, **81**, 1266.

80. Prichard, W. W., U.S.P. 2, 769, 003; Chem. Abs., 1957, **51**, 7412.

81. Prichard, W. W., U.S.P. 2, 841, 591; Chem. Abs., 1958, **52**, 20197.

82. Reppe, W., Annalen, 1953, **582**, 1.

83. Reppe, W. and Kroper, H., Annalen, 1953, **582**, 38.

84. Reppe, W. and Vetter, H., Annalen, 1953, **582**, 133.

85. Rhone-Poulenc, S. A., Fr. Addn., 84, 383; Chem. Abs., 1965, **63**, 1716.

86. Rosenthal, A., Canad. J. Chem., 1960, **38**, 2025.

87. Rosenthal, A., J. Org. Chem., 1961, **26**, 1638.

88. Rosenthal, A., Astbury, R. F. and Hubscher, A., J. Org. Chem., 1960, **25**, 1037.

89. Rosenthal, A. and Gervay, J., Chem. and Ind., 1963, 1623.

90. Rosenthal, A. and Gervay, J., Canad. J. Chem., 1964, **42**, 1490.

91. Rosenthal, A. and Hubscher, A., J. Org. Chem., 1960, **25**, 1562.

92. Rosenthal, A. and Millward, S., Canad. J. Chem., 1963, **41**, 2504.

93. Rosenthal, A. and Millward, S., Canad. J. Chem., 1964, **42**, 956.

94. Rosenthal, A. and O. Donnell, J.P., Canad. J. Chem., 1960, **38**, 457.

95. Rosenthal, A. and Weir, M. R. S., Canad. J. Chem., 1962, **40**, 610.

96. Rosenthal, A. and Weir, M. R. S., J. Org. Chem., 1963, **28**, 3025.

97. Rosenthal, A. and Yalpani, M., Canad. J. Chem., 1965, **43**, 711.

98. Rosenthal, A. and Yalpani, M., Canad. J. Chem., 1965, **43**, 3449.

99. Ryang, M., Myeong, S. K., Sawa, Y. and Tsutsumi, S., J. Organo-metallic Chem., 1966, 5, 305.

100. Ryang, M., Rhee, I. and Tsutsumi, S., Bull. Chem. Soc. Japan, 1964, **37**, 341.

101. Ryang, M., Sawa, Y., Masada, H. and Tsutsumi, S., J. Chem. Soc. Japan, Ind. Chem. Sect., 1963, **66**, 1086.

102. Ryang, M. and Tsutsumi, S., Bull. Chem. Soc. Japan,, 1962, **35**, 1121.

Transition Metal Intermediates

103. Sampson, H. J., U.S.P., 2, 589, 289; Chem. Abs., 1952, **46**, 11234.

104. Sauer, J. C., Cramer, R. D., Engelhardt, V. A., Ford, T. A., Holmquist, H. E. and Howk, B. W., J. Amer. Chem. Soc., 1959, **81**, 3677.

105. Sauer, J. C., Howk, B. W. and Stiehl, R. T., J. Amer. Chem. Soc., 1959, **81**, 693.

106. Schrauzer, G. N., Chem. and Ind., 1958, 1404.

107. Schrauzer, G. N., J. Amer. Chem. Soc., 1959, **81**, 5307.

108. Stern, E. W. and Spector, M. L., J. Org. Chem., 1966, **31**, 596.

109. Sternberg, H. W., Markby, R. and Wender, I., J. Amer. Chem. Soc., 1958, **80**, 1009.

110. Sternberg, H. W., Wender, I., Friedel, R. A. and Orchin, M., J. Amer. Chem. Soc., 1953, **75**, 3148.

111. Strohmeier, W., Laporte, H. and Hobe, D., Chem. Ber., 1962, **95**, 455.

112. Tate, D. P., Augl, J. M., Ritchey, W. M., Ross, B. L. and Grasselli, J. G., J. Amer. Chem. Soc., 1964, **86**, 3261.

113. Tsuji, J. and Iwamoto, N., Chem. Comm., 1966, 380.

114. Vallarino, L. M. and Santarella, G., Gazzetta, 1964, **94**, 252.

115. Weiss, E. and Hübel, W., J. Inorg. Nuclear Chem., 1959, **11**, 43.

116. Weiss, E., Hübel, W. and Merényi, R., Chem. Ber., 1962, **95**, 1155.

117. Wender, I., Friedman, S., Steiner, W.A. and Anderson, R. B., Chem. and Ind., 1958, 1694.

CHAPTER 9

DECARBONYLATION AND DESULPHONYLATION REACTIONS

The reactions to be considered here are, in effect, the converse of those considered in the preceding chapters. The ability of transition metals to catalyse decarbonylation reactions appears to have been observed first by Sabatier and Senderens[22] and antedates the discovery of carbonylation reactions. Much of the published work relates to the use of metallic catalysts, but more recently complexes have been found which also catalyse the reaction. In view of the close similarity between the two types of catalyst, the former kind has been included in this discussion despite its heterogeneous character.

(1)

The metallic catalyst most frequently employed is palladium on charcoal or barium sulphate. The latter appears to diminish the incidence of accompanying hydrogen transfer and dehydrogenation reactions. Such reactions are typified by the conversion of the aldehyde (1) to naphthalene by heating with palladium on charcoal[18]. Decarbonylation reactions are generally limited to aldehydes and acid halides, which are converted to the corresponding hydrocarbons or halides respectively (Table 9.1).

$$R.CHO \longrightarrow RH + CO$$

$$R.CO.Cl \longrightarrow R.Cl + CO$$

Decarbonylation of *trans*-α-substituted cinnamaldehydes (2) proceeds stereospecifically to the *cis*-olefin (3)[11]. Isomerisation to the *trans*-olefin (4) occurs independently.

| (2) | (3) | (4) |

The rate of decarbonylation of these cinnamaldehydes depends on the nature of R and decreases in the order: hydrogen > phenyl > methyl > ethyl > n-propyl > iso-propyl[12]. The direction of the

TABLE 9.1 Decarbonylation of Aldeydes and Acid Halides

Aldehyde or Acid Halide	Catalyst *	Product	Yield(%)	References
Butyraldehyde	A	Propane	—	24
iso-Butyraldehyde	A	Propane	—	24
Citral	B	Geraniolene	—	7
Citronellal	B	Citronellene	—	7
Myrtenal	B	Apopinene	73	8
Benzaldehyde	A	Benzene	83	24
	B	Benzene	—	7
	C	Benzene	78	10
p-Tolualdehyde	C	Toluene	88	10
p-Chlorobenzaldehyde	A	Chlorobenzene	85	24
o-Methoxybenzaldehyde	C	Anisole	94	10
Piperonal	D	Phenol and catechol	—	14
p-Hydroxybenzaldehyde	A	Phenol	70	24
Vanillin	D	o-Methoxyphenol and catechol	—	14
m-Nitrobenzaldehyde	C	Nitrobenzene	86	10
p-Nitrobenzaldehyde	C	Nitrobenzene	89	10
1-Naphthaldehyde	C	Naphthalene	80	10
1, 2, 3, 4-Tetrahydro-2-naphthaldehyde	C	Naphthalene	97	18
5, 6, 7, 8-Tetrahydro-2-naphthaldehyde	C	Tetralin and naphthalene	—	19
1-Formylfluorenone	C	Fluorenone	82	10

Substrate	Method	Product	Yield (%)	Ref.
9-Anthraldehyde	C	Anthracene	84	10
Biphenyl-2,2'-dicarboxaldehyde	C	Biphenyl	97	10
Methyl 2-formylbiphenyl-2'-carboxylate	C	Methyl biphenyl-2-carboxylate	20	10
Dihydrocinnamaldehyde	A	Ethylbenzene	67	24
Cinnamaldehyde	A	Styrene	77	24
	D	Styrene, ethylbenzene, toluene, benzene	—	14
trans-α-Methylcinnamaldehyde	C	β-Methylstyrene (85% cis)	92	11
trans-α-Ethylcinnamaldehyde	C	β-Ethylstyrene (61% cis)	85	11
trans-α-n-Propylcinnamaldehyde	C	β-n-Propylstyrene (cis and trans) and n-Pentylbenzene	—	11
trans-α-iso-Propylcinnamaldehyde	C	β-iso-Propylstyrene (cis and trans)	48	11
trans-α-Phenylcinnamaldehyde	C	Stilbene (64% cis)	42	11
Furfural	B	Furan	—	7
	C	Furan	94	15
Pyridine-2-aldehyde	C	Pyridine	68	10
Octanoyl bromide	A	Hept-1-ene (71%), hept-2 and 3-enes (29%)	91	25a
Decanoyl chloride	E	Heptenes	80	26
	C	Nonenes	58	26
	E	Nonenes	90	26
Adipoyl chloride	E	Cyclopentanone	30	26
Phenylacetyl chloride	A	Benzyl chloride	86	25, 25a
	C	Benzyl chloride	42	26

(Continued overleaf)

TABLE 9.1 (continued)

Aldehyde or Acid Halide	Catalyst *	Product	Yield(%)	References
1-Naphthylacetyl chloride	A	1-Chloromethylnapthalene	87	4
Dihydrocinnamoyl chloride	A	Styrene	71	25
	E	Styrene	53	26
Benzoyl chloride	A	Chlorobenzene	85	25a
	C	Chlorobenzene	75	13, 17, 26
	D	Chlorobenzene	50	13, 17
o-Bromobenzoyl chloride	A	o-Bromochlorobenzene	78	4
p-Chlorobenzoyl chloride	A	p-Dichlorobenzene	79	4
2,4-Dichlorobenzoyl chloride	A	1,2,4-Trichlorobenzene	98	4
o-Iodobenzoyl chloride	A	o-Iodochlorobenzene	38	4
p-Iodobenzoyl chloride	A	p-Iodochlorobenzene	78	4
Isophthaloyl chloride	C	m-Dichlorobenzene	68	13, 17
1-Naphthoyl chloride	A	1-Chloronaphthalene	96	4
2-Methyl-1-naphthoyl chloride	A	1-Chloro-2-methylnaphthalene	93	4
2-Naphthoyl chloride	A	2-Chloronaphthalene	94	4

* A, $(Ph_3P)_3RhCl$; B, $Pd/BaSO_4$; C, Pd/C; D, Ni; D, $PdCl_2$

242

structural effect suggests that steric factors are concerned, but the position of the phenyl group in this series must be due to electronic effects.

The ability of acid halides to undergo decarbonylation with palladium catalysts emphasises the relationship of this reaction to the Rosenmund reduction, where acylpalladium intermediates, R.CO.PdCl, are probably involved[6, 26]. Heating decanoyl chloride with either palladium on charcoal or palladous chloride, which is decomposed to palladium, yields nonenes[26]. This reaction is obviously the converse of the palladium catalysed hydrocarboxylation of olefins described in Chapter 7. Similar olefin formation occurs with octanoyl bromide and β-phenylpropionoyl chloride. Adipoyl chloride yields cyclopentanone, possibly by way of pent-4-enoyl chloride (5).

$$(CH_2)_4 \begin{array}{l} CO \cdot Cl \\ CO \cdot Cl \end{array} \longrightarrow \begin{array}{l} CH{=}CH_2 \\ CH_2 \\ CH_2 \cdot CO \cdot Cl \end{array} \longrightarrow \bigcirc{=}O$$

(5)

The cyclisation presumably involves an acylpalladium species and parallels similar conversions described in Chapter 8. In the case of decarbonylation of aldehydes the intermediate complex, R.CO.PdH, is the same as that involved in the Rosenmund reduction, where at lower temperatures it decomposes to aldehyde and palladium.

The versatile complex $(Ph_3P)_3RhCl$, which reversibly dissociates in solution to $(Ph_3P)_2RhCl$, also catalyses the decarbonylation of aldehydes[24] and acid halides[4]. In the case of aldehydes the products are the hydrocarbon and $[(Ph_3P)_2Rh(CO)Cl]$ at room temperature or in refluxing benzene[1, 24]. Kinetic studies indicate that the reaction is first order in aldehyde and in the complex[1]. Use of this reagent has made possible a facile synthesis of 1-dehydro-3-keto-steroids of the A/B *cis* series, the key step being the decarbonylation of (6) to (7) in 68% yield[23]. As pointed out, the method is also applicable to the introduction of a double bond on the methylene side of ketones of the type $R^1R^2CH.CH_2COCHR^3CH_2R^4$ especially in cases where enolisation towards the methine carbon is preferred.

(6) **(7)**

Transition Metal Intermediates

The reaction of palmitoyl chloride with $[(Ph_3P)_3RhCl]$ in boiling benzene yields the complex $[CH_3(CH_2)_{14}CO.Rh(Ph_3P)_2Cl_2]$[25], whereas under similar conditons benzoyl chloride yields $[C_6H_5Rh(Ph_3P)_2(CO)Cl_2]$[1,25a]. Acetyl chloride gives methyl chloride and $[(Ph_3P)_2Rh(CO)Cl]$, but under milder conditions the complex $[CH_3Rh(Ph_3P)_2(CO)Cl_2]$ can be isolated. These observations suggest the sequence

$$(Ph_3P)_3RhCl \rightleftharpoons (Ph_3P)_2RhCl \xrightarrow{R\,CO\,Cl} R.CO.Rh(Ph_3P)_2Cl_2$$

$$\rightleftharpoons R.Rh(Ph_3P)_2(CO)Cl_2 \nearrow^{\text{olefin + HCl} + (Ph_3P)_2Rh(CO)Cl}_{\searrow RCl + (Ph_3P)_2Rh(CO)Cl}$$

$$\textbf{(8)} \qquad\qquad\qquad\qquad \textbf{(9)}$$

At a higher temperature, 200°C, the reaction becomes catalytic as the complex $[(Ph_3P)_2Rh(CO)Cl]$ is decomposed[4]. The reversible nature of these reactions is illustrated by the carbonylation of benzyl chloride to phenylacetyl chloride in benzene at 150°C and 100 atmospheres in the presence of complex (9)[25a]. The complex $(8, R=C_6H_5.CH_2)$ is formed as an intermediate. Decarbonylation of alkanoyl halides with $[(Ph_3P)_3Rh\,Cl]$ gives predominantly the terminal olefin, whereas mostly internal olefins are obtained from the palladium catalysed reaction.

Carboxylic acids are converted into olefins by heating under reflux with the complex $[(Et_2PhP)_3RhCl_3]$,[21] which is finally recovered as $[(Et_2PhP)_2Rh(CO)Cl]$. Propionic acid is thus converted into ethylene and hexanoic acid into pent-2-ene, up to three molecules of olefin being formed per atom of rhodium.

The complex $[Ru_2Cl_3(Et_2PhP)_6]^+Cl^-$ converts butyraldehyde into propylene and propane, and is itself converted into $[(Et_2PhP)_3Ru(CO)Cl_2]$[21]. The first stage in the reaction is the formation of $[(Et_2PhP)_3RuCl_2]_n$. Analogous triphenylphoshpine complexes have been reported to dissociate like the foregoing rhodium complex, and also to catalyse hydrogenation[9].

Decarbonylation of compounds other than acid halides and aldehydes have rarely have been reported. Diphenylcyclopropenone is decarbonylated to diphenylacetylene by nickel, iron and cobalt carbonyls[2,3]. Similarly, dibutylcyclopropenone yields dibutylacetylene. Urea is decarbonylated to hydrazine by heating with iron, cobalt, nickel, molybdenum or tungsten metal[20]. A decarbonylation-carbonylation process is presumably involved in the exchange of ^{14}CO with dicobalt octacarbonyl at 230°C.[27] Similar exchanges occur with acetic anhydride and phthalic anhydride. In the presence of hydrogen the latter compound is converted into benzoic acid.

244

TABLE 9.2 Desulfonylation of Arylsuphonyl Halides

Sulphonyl Halide	Catalyst*	Product	Yield(%)	References
Benzenesulphonyl Chloride	A	Chlorobenzene	79	5
	F	Chlorobenzene	91	16
Benzenesulphonyl bromide	F	Bromobenzene	—	16
p-Toluensulphonyl chloride	A	p-Chlorotoluene	72	5
p-Chlorobenzenesulphonyl chloride	A	p-Dichlorobenzene	85	5
2,5-Dichlorobenzenesulphonyl chloride	A	1,2,4-Trichlorobenzene	65	5
p-Fluorobenzenesulphonyl chloride	A	p-Chlorofluorobenzene	61	5
Benzene-1,3-disulphonyl chloride	A	m-Dichlorobenzene	62	5
	F	m-Dichlorobenzene	—	16
4-Fluoronaphthalene-1-sulphonyl chloride	A	4-Chloro-1-fluoronaphthalene	70	5
Naphthalene-2-sulphonyl chloride	A	2-Chloronaphthalene	69	5
Naphthalene-1,5-disulphonyl chloride	F	1,5-Dichloronaphthalene	—	16
Naphthalene-2,7-disulphonyl chloride	F	2,7-Dichloronaphthalene	—	16

* A, $(Ph_3P)_3RhCl$; F, CuCl

Transition Metal Intermediates

Phthalide, anthraquinone, phenanthraquinone and benzophenone do not undergo exchange under similar conditions.

In view of the ability of sulphur dioxide to undergo insertion into carbon-metal bonds it is not surprising that sulphonyl halides can undergo desulphonylation to halides. (Table 9.2). Again, $[(Ph_3P)_3RhCl]$[5] can be used as the catalyst and intermediates analogous to those involved in decarbonylation of acid halides are presumably involved. The ability of palladium to catalyse this reaction is claimed[16] but another report[5] suggests that this is incorrect. The use of cuprous chloride has also been reported[16]. The threshold temperature for desulphonylation with the rhodium complex is 235°C, whereas that for decarbonylation is 190°C.

REFERENCES

1. Baird, M. C., Lawson, D. N., Mague, J. T., Osborn, J. A. and Wilkinson, G., Chem. Comm., 1966, 129.

2. Bird, C. W. and Hollins, E. M., unpublished observations.

3. Bird, C. W. and Hudec, J., Chem. and Ind., 1959, 570.

4. Blum, J., Tetrahedron Letters, 1966, 1605.

5. Blum, J., Tetrahedron Letters, 1966, 3041.

6. Chiusoli, G. P. and Agnès, G., Chimica e Industria, 1964, **46**, 548.

7. Eschinazi, H. E., Bull. Soc. Chim. France, 1952, 967.

8. Eschinazi, H. E., J. Org. Chem., 1959, **24**, 1369.

9. Evans, D., Osborn, J. A., Jardine, F. H. and Wilkinson, G., Nature, 1965, **208**, 1203.

10. Hawthorn, J. O. and Wilt, M. H., J. Org. Chem., 1960, **25**, 2215.

11. Hoffman, N. E., Kanakkanatt, A. T. and Schneider, R. F., J. Org. Chem., 1962, **27**, 2687.

12. Hoffman, N. E. and Puthenpurackal, T., J. Org. Chem., 1965, **30**. 420.

13. McCall. E. B. and Bain, P. J. S., B. P. 957, 957; Chem. Abs., 1964, **61**, 5563.

14. Mailhe, A., Bull. Soc. Chim. France, 1926, **39**, 922.

15. Marukawa, K., Takanaka, Y., Kubota, T. and Sato, H., Jap. P., 1531, 1532, 1533 ('65); Chem. Abs., 1966, **64**, 12644.

16. Monsanto Chemicals Ltd., Fr. P. 1, 340, 883; Chem. Abs., 1964, **60**, 5393.

17. Monsanto Chemicals Ltd., Fr. P. Addn. 83, 657; Chem. Abs., 1965, **62**, 16117.

18. Newman, M. S. and Mangham, J. R., J. Amer. Chem. Soc., 1949, **71**, 3342.

19. Newman, M. S. and Zahm, H. V., J. Amer. Chem. Soc., 1943, **65**, 1097.

20. Passino, H. J., U. S. P. 2, 675, 301; 2, 717, 201. Chem. Abs., 1954, **48**, 9636; 1956, **50**, 2131.

21. Prince, R. H. and Raspin, K. A., Chem. Comm., 1966, 156.

22. Sabatier, P. and Senderens, J. B., Ann. Chim. Phys., 1905, **4**, 433.

23. Shimizu, Y., Mitsuhashi, H. and Caspi, E., Tetrahedron Letters, 1966, 4113.

24. Tsuji, J. and Ohno, K., Tetrahedron Letters, 1965, 3969.

25. Tsuji, J. and Ohno, K., J. Amer. Chem. Soc., 1966, **88**, 3452.

25a. Tsuji, J. and Ohno, K., Tetrahedron Letters, 1966, 4713.

26. Tsuji, J. Ohno, K., and Kajimoto, Y., Tetrahedron Letters, 1965, 4565.

27. Wender, I., Friedman, S., Steiner, W. A. and Anderson, R. B., Chem. and Ind., 1958, 1694.

CHAPTER 10

HOMOGENEOUS HYDROGENATION

Homogeneous hydrogenation catalysts have a number of advantages over the heterogeneous ones conventionally used. In particular the problems of reproducibility of catalyst behaviour and the extensive modification, or even poisoning, of catalytic activity by stray contaminants are largely avoided.

Since the first observation of the homogeneous reduction of quinoline solutions of benzoquinone by hydrogen in the presence of cupric acetate at 100°C[3], a variety of other homogeneous hydrogenation systems have been found.

The first stage in this hydrogenation is the reduction of copper (II) to copper (I). Subsequent reaction of copper (I) with hydrogen generates copper hydride which is apparently the reducing agent. The catalytic activity is affected by the solvent[87]. Dodecylamine, pyridine and quinoline derivatives permit catalytic activity but dimethylaniline, dipentylamine, diethanolamine, formamide, 8-hydroxyquinoline and dibutyl phthalate do not. Ethylenediamine and ethylenediaminetetra-acetic acid inhibit catalytic activity[90]. In quinoline solution cuprous acetate, salicylaldehyde and 4-hydroxysalicylaldehyde have comparable catalytic activities, which are somewhat higher than those of the stearate and benzoate[4]. However, nitrobenzoates and nitrosalicylaldehydes of copper (I) as well as various Schiff's bases are inactive. At higher temperatures, 280°C, oleic and palmitic acids are reduced to the corresponding saturated alcohols, but inclusion of a cadmium salt inhibits reduction of the double bond and good yields of oleoyl alcohol are produced[70].

Salts of chromium (III)[77], manganese (II)[76], iron (III)[75, 82,] cobalt (II)[75, 80], nickel (II)[75, 78] and copper (II)[81] catalyse the hydrogenation of cyclohexene and other olefins at temperatures of 20 to 60°C and pressures up to 100 atmospheres. Water acts as a catalyst poison. Magnetic measurements indicate a strong interaction of cyclohexene with the catalysts[79]. The partial order of decreasing catalytic activity is iron (III) > cobalt (II) > nickel (II)[75]. A different order of reactivity has been found using acetylacetonate complexes for the hydrogenation of linseed and soybean methyl esters, which contain large proportions of linoleic and linolenic acids[7]. The order observed is nickel (II) > cobalt (III) > copper (II) > iron (III). The hydrogenation occurs rapidly in methanol at 100–180°C and 7 to 70 atmospheres, but more slowly in acetic acid or dimethylformamide. The triene and diene esters are preferentially reduced to mono-unsaturated esters.

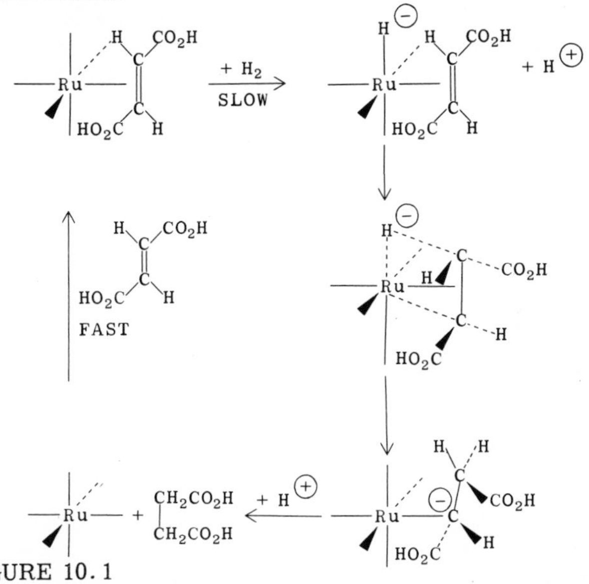

FIGURE 10.1

Ruthenium (II) in aqueous hydrochloric acid catalyses the homogeneous hydrogenation of maleic, fumaric and acrylic acid[18]. The formation of 1:1 olefin-ruthenium (II) complexes under these conditions has been observed. The observed rate law is of the form $k[H_2][Ru^{II}$-olefin$]$. Reduction of fumaric acid with deuterium in aqueous solution yields undeuterated succinic acid, whereas reduction with hydrogen or deuterium in deuterium oxide gives dl-2,3-dideuterosuccinic acid. Thus the hydrogen which adds in a cis-manner to the double bond is derived from the solvent. This catalyst does not promote reduction of isolated double-bonds although ruthenium (II) complexes are formed, which catalyse exchange between deuterium and water at about the same rate as the hydrogenation of fumaric acid. The mechanism summarised in Figure 10.1 has been proposed for this hydrogenation. A closely parallel system is provided by the triphenylphosphine complexes $[(Ph_3P)_4RuCl_2]$ and $[(Ph_3P)_3RuCl_2]$[8]. Molecular weight studies in benzene show the existence of the following equilibria:

$$RuCl_2(Ph_3P)_4 \rightleftarrows RuCl_2(Ph_3P)_3S + Ph_3P$$

$$RuCl_2(Ph_3P)_3 \rightleftarrows RuCl_2(Ph_3P)_2S_2 + Ph_3P$$

Transition Metal Intermediates

Treatment of concentrated solutions of the complex in
ethanol: benzene (1:1) with hydrogen generates the hydrido complex
$[(Ph_3P)_3RuClH]$. This system catalyses the hydrogenation of hept-
1-ene and hex-1-yne at 25°C and 1 atmosphere in a benzene-
ethanol (1:1) solvent. The hydrogenation is extremely slow in the
absence of ethanol showing that it plays an important rôle. Mixed
hydrogen deutero paraffins are formed when deuterium is used.
In the absence of olefin rapid exchange occurs between hydroxyl
O–H and deuterium.

Rhodium (III) and rhodium (I) chlorides are not effective catalysts
for homogeneous hydrogenation of maleic acid in aqueous solution[23].
In ethanolic solution simple olefins such as hex-1-ene are reduced
at atmospheric pressure in the presence of rhodium trichloride,
1, 2, 6-trichlorotripyridinerhodium (III), 1, 2, 3, tris-(triphenylphosphine) trichlororhodium (III) or its dimethylphenylarsine
analogue[13, 58]. It has been suggested that a hydride species, e.g.
$(C_6H_5N)_3RhCl_2H$, is involved. Rhodium trichloride in dimethylacetamide solution catalyses the reduction of maleic acid[23]. The
rhodium (III) is reduced by hydrogen to rhodium (I), and then forms
a complex with maleic acid which is then reduced to succinic acid.
The solvated rhodium (I) chlorine-bridged stannous chloride complex $[Rh_2Cl_2(SnCl_2, C_2H_5OH)_4]$ catalyses the hydrogenation of hex-
1-ene in ethanolic solution[58]. The connection between hydride formation and homogeneous hydrogenation has been demonstrated with
the complex $(Ph_3P)_3RhCl$,[91] which in ethanol-benzene solution
catalyses the rapid reduction of olefins and acetylenes at 25°C and
1 atmosphere. This complex dissociates in solvents (S) forming
$[(Ph_3P)_2RhCl(S)]$, which readily takes up molecular hydrogen forming $[(Ph_3P)_2RhCl(S)H_2]$. Addition of olefin causes instantaneous reduction of the olefin, and subsequent introduction of more molecular
hydrogen regenerates the hydride complex. So far no evidence has
been obtained for the formation of substantial amounts of an alkyl-
rhodium intermediate.

Comparable behaviour has been observed with iridium complexes[83,84]. Square-planar complexes of iridium (I), $[IrX(CO)(Ph_3P)_2]$ X = halogen, react reversibly with hydrogen and ethylene[84].

Similar reversible uptakes of hydrogen and ethylene have been observed with $[(Ph_3P)_3 Ir H(CO)]$[83]. These iridium complexes catalyse the hydrogenation of ethylene and acetylene. The corresponding rhodium and osmium complexes display similar catalytic activity but do not take up hydrogen although exchange does occur.

$$Rh\ H(CO)(Ph_3P)_3 \xrightarrow{D_2} RhD\ (CO)(Ph_3P)_3$$

$$OsHCl(CO)(Ph_3P)_3 \xrightarrow{D_2} Os\ D\ Cl(CO)(Ph_3P)_3$$

In the latter case an 8-coordinate intermediate is apparently involved. A possible mechanism for the hydrogenation of ethylene is

$$L_4\ Ir\ H + C_2H_4 \rightleftharpoons L_4\ Ir\ H(C_2H_4) \rightleftharpoons L_4\ Ir(C_2H_5)$$

$$L_4\ Ir(C_2H_5) + H_2 \rightleftharpoons L_4\ Ir\ H_2(C_2H_5) \longrightarrow L_4\ IrH + C_2H_6$$

where $L_4 = CO(Ph_3P)_3$.[83]

A platinum (II) stannous chloride catalyst has been found to catalyse the homogeneous hydrogenation of ethylene and acetylene in methanolic solution[5]. The catalyst contains $SnCl_3^-$ ligands bonded to platinum via tin. Higher olefins are less readily hydrogenated, which correlates with their lower ability to complex with platinum. The homogeneous hydrogenation of ethyleneplatinous chloride has been reported[9].

Palladous chloride has also been shown to catalyse the reduction of ethyl crotonate in aqueous ethanolic solution[39]. Salts of copper, nickel, zinc, cobalt, silver, mercury, cadmium, sodium, calcium, aluminium, magnesium, cerium and chromium, promote the catalytic activity. Sodium, the most effective promoter, increases the activity by 6-7 times. Although palladium metal is deposited during the reduction it was found not to be catalytically active.

In view of the foregoing survey of transition metal ions which show catalytic activity in homogeneous hydrogenation reactions a comparative survey of their activities for catalysis of the reduction of dicyclopentadiene is of especial interest. In dimethylformamide or dimethylacetamide at 25°C and 1 atmosphere the order of decreasing activity is palladous chloride > rhodium trichloride ~ ruthenium trichloride > potassium tetrachloroplatinate[60]. Addition of thiophen to the palladous chloride system markedly increases the activity. 1, 2- and 1, 4-Naphthoquinones are also reduced under these conditions.

Transition Metal Intermediates

One of the most extensively investigated catalysts of homogeneous hydrogenation is the pentacyanocobaltate (II) anion. Solutions of the pentacyanocobaltate (II) anion are conveniently obtained by the addition of a cyanide solution to a cobalt (II) salt. These solutions react with molecular hydrogen, sodium borohydride or hydrazine forming the hydridopentacyanocobaltate (III) anion[17,22,25,86]. It seems that the reaction with hydrogen entails homolytic splitting of the hydrogen molecule[41,65]. The same hydrido complex, as well as the corresponding hydroxo complex, is also formed, by the "aging" of aqueous solutions of pentacyanocobaltate (II) anion on standing, as a result of the homolytic cleavage of water.

$$2[Co(CN)_5]^{3-} + H_2 \rightleftharpoons 2[Co(CN)_5H]^{3-}$$
$$2[Co(CN)_5]^{3-} + H_2O \rightleftharpoons [Co(CN)_5H]^{3-} + [Co(CN)_5OH]^{3-}$$

An aqueous or semiaqueous medium is thus used for the hydrogenation reactions and it is frequently necessary to keep it strongly basic. Although many reductions can be effected at room temperature and atmospheric pressure, higher temperatures, 40-90°C, and pressures up to 100 atmospheres have been used.

Carbon-carbon double and triple bonds are only reduced when part of a conjugated system. However, as Table 10.1 shows, there are some puzzling inconsistencies. Acrylic, crotonic and tiglic acids are not reduced under conditions which effect hydrogenation of methacylic, itaconic and atropic acids. At higher temperatures acrylic acid gives propionic and α-methylglutaric acids. The latter is probably derived from 3-methylglutaconic acid which is formed in the absence of hydrogen[30]. However, it is noteworthy that ethyl acrylate and acrylonitrile form adipic acid[50]. The position with regard to the reduction of $\alpha\beta$-unsaturated aldehydes is almost the converse of that found with the corresponding acids. Tiglic aldehyde, and crotonaldehyde are hydrogenated but acrolein and methacrolein are not. In view of the basic medium employed it is not surprising that in some cases condensation products have been obtained from aldehydes. The reduction of diphenylmaleic anhydride yields *dl*-diphenyl-succinic acid showing that an overall *trans* addition of hydrogen has occurred.

The reduction of nitro-compounds generally leads to bimolecular products. It is essential in the case of nitro-compounds to add the compound slowly to the catalyst solution so as to ensure the presence of an excess of catalyst throughout the reaction.

Oximes are reduced to amines. Cobalt complexes of α-oximinocarboxylic acids can be hydrogenated to amino-acids under apparently homogeneous conditions[24,47,48,49]. The yields are very poor, unless cyanide ion is added. Other ions such as chloride and

252

TABLE 10.1 Hydrogenations Catalysed by Pentacyanocobaltate (II)

Substrate	Products	References
Butadiene	But-1-ene and/or *trans*-but-2-ene	30, 71, 72
Isoprene	2-Methylbut-1-ene or 2-methylbut-3-ene	29, 30, 73
Cyclopentadiene	Cyclopentene	30
Cyclohexa-1,3-diene	Cyclohexene	29, 30
Styrene	Ethylbenzene	29, 30
α-Methylstyrene	Isopropylbenzene	30
Propenylbenzene	Not reduced	29, 30
Indene	Not reduced	29
1,1-Diphenylethylene	Not reduced	29, 30
Stilbene	Not reduced	30
Allyl chloride	Propylene	29
Allyl acetate	Propylene	29
Diphenylmethyl bromide	1,1,2,2-Tetraphenylethane	31
Triphenylmethyl bromide	Hexaphenylethane	31

(Continued overleaf)

TABLE 10.1 (continued)

Substrate	Products	References
Tropylium bromide	Bitropyl	31
Cinnamyl alcohol	3-Phenylpropanol	30
Cyclohexene oxide	Cyclohexanol	29
Styrene oxide	β-Phenylethanol	29
Acrolein	Not reduced	30
Methacrolein	Not reduced	30
Crotonaldehyde	n-Butyraldehyde	30
Tiglic aldehyde	α-Methylbutyraldehyde	26, 29, 30
α-Methylpentenal	α-Methylvaleraldehyde	26, 30
Propionaldol	α-Methylvaleraldehyde	30
Propionaldehyde	α-Methylvaleraldehyde	29, 30
Benzaldehyde	Benzyl alcohol	27, 30
Benzaldehyde and acetaldehyde	α-Benzylcinnamaldehyde	30
Cinnamaldehyde	2-Benzyl-5-phenylpentenal	29, 30
Methyl isopropenyl ketone	Methyl isopropyl ketone	26
Benzylideneacetone	1-Phenylbutan-3-one	28

1,2-Dibenzoylethane	28
Hydroquinone	27, 29, 30
9,10-Dihydroxyanthracene	30
erythro-1,2-Diphenylethanolamine	52
Benzoin	29
Benzoin, *erythro*-1,2-diphenylethanolamine, *erythro*-1,2-diphenylethylenediamine	52
Propionic acid, α-methylglutaric acid	29, 50
Isobutyric acid	29, 30
Not reduced	30
Not reduced	30
β-Phenylpropionic acid	29, 30, 46, 51, 66
α-Phenylpropionic acid	30
Hex-2,3 and 4-enoic acids	29, 36, 37, 46, 51, 85
Succinic acid	52
Fumaric and succinic acids	51, 52
dl-Diphenylsuccinic acid	51, 52
α-Methylsuccinic acid	26, 30

Dibenzoylethylene
Benzoquinone
9,10-Anthraquinone
Benzoin and ammonia
Benzil
Benzil and ammonia

Acrylic acid
Methacrylic acid
Crotonic acid
Tiglic acid
Cinnamic acid

Atropic acid
Sorbic acid

Maleic acid
Acetylenedicarboxylic acid
Diphenylmaleic anhydride
Itaconic acid

(Continued overleaf)

TABLE 10. 1 (continued)

Substrate	Products	References
Ethyl acrylate	Adipic acid	50
Methyl methacrylate	Isobutyric acid	26, 29
Pyruvic acid and ammonia	Alanine	45, 46
Phenylpyruvic acid and ammonia	Phenylalanine	45, 46
α-Ketoglutaric acid and ammonia	α-Aminoglutaric acid	45
Ethyl pyruvate and ammonia	Alanine	46
Ethyl phenylpyruvate and ammonia	Phenylalanine	46
Phenylacetonitrile	Phenylacetic acid	50
Acrylonitrile	Adipic acid	50
Cinnamonitrile	β-Phenylpropionic acid and amide	50
Benzoyl cyanide	Benzoic acid	50
Acetoxime	iso-Propylamine	46
Acetophenone oxime	α-Phenylethylamine	46
Phenylpyruvic acid oxime	Phenylalanine	46
Benzoyl cyanide oxime	α-Aminophenylacetic acid, benzoic acid	50
Phenylacetyl cyanide phenylhydrazone	2-(N'-Phenylhydrazino)-3-phenylpropionic acid	50

N-Nitrosodimethylamine	1,1-Dimethylhydrazine, dimethylamine	27
Nitroethane	Ethylamine, ethanol, acetaldehyde	50
2-Nitro-2-methylpropane	t.Butylamine	50
Nitrobenzene	Aniline, azoxybenzene	22
	Azobenzene	29, 50
	Hydrazobenzene	50
	Azobenzene, hydrazobenzene	27, 30, 50
o-Nitrotoluene	2,2'-Dimethylazobenzene, 2,2'-dimethylazoxybenzene	30
p-Nitrotoluene	4,4'-Dimethylazoxybenzene, p-tolylhydroxylamine	27, 30
o-Chloronitrobenzene	2,2'-Dichloroazoxybenzene, 2,2'-dichloroazobenzene, 2,2'-dichlorohydrazobenzene	27
o-Nitrophenol	o-Aminophenol	50
m-Nitrophenol	No reduction	50
p-Nitrophenol	p-Aminophenol	50
o-Nitroanisole	2,2'-Dimethoxyhydrazobenzene	30
Azoxybenzene	Azobenzene	29

thiocyanate are far less effective[49]. The best yield, 49%, of α-alanine from bis(pyruvic acid oximato) cobalt (II) is obtained at 65-70°C and 100 atmospheres[48]. Lower temperatures result in lower yields but these can be improved by addition of alkali. The optimum molar ratio of cyanide ion to complex is 5.5[24,48]. The experimental data suggest that the catalyst is a species in which only part of the ligands is replaced by cyanide anions. Alternatively α-aminoacids can be obtained by reductive amination of α-ketoacids in the presence of ammonia and pentacyanocobaltate (II)[45,46]. The reductive amination of benzoin and benzil produces the erythro amino-alcohol, showing that hydrogen addition occurs at the less sterically hindered side of the imine intermediate[52].

The butene obtained by reduction of butadiene depends on the cyanide to cobalt ratio in the catalyst solution[30,71].

Molar ratio CN^-/Co^{2+}	trans-but-2-ene	cis-but-2-ene	but-1-ene
4.5	86%	1	13
5.5	70	1	29
6.0	12	3	85
8.5	19	1	80

The presence of both $[Co(CN)_3]^-$ and $[Co(CN)_4]^{2-}$ has been detected at CN^-/Co^{2+} ratios below 5[74]. When potassium hydroxide is added to the catalyst solution the threshold value is shifted to between 4.9 and 5.0, when the complex is hydrogenated before addition of butadiene[71]. The order of addition of potassium hydroxide and cyanide to cobaltous ion also determines the major product[72]. Catalyst solutions prepared by successive addition of potassium hydroxide and potassium cyanide effect the reduction of butadiene to but-1-ene. The converse order of addition results in the formation of trans-but-2-ene. Similarly, reduction of isoprene yields over 90% of 2-methylbut-2-ene when the CN/Co ratio is below 5.1[73]. At higher ratios the major product is 2-methylbut-3-ene with lesser amounts of 2-methylbut-1- and 2-enes.

The information so far available suggests that the mechanism of pentacyanocobaltate (II) catalysed hydrogenation depends on the substrate. Complexes of the type $[Co(CN)_5R]^{3-}$, which have been postulated as intermediates in these reductions[31], can be prepared either by addition of $[Co(CN)_5H]^{3-}$ to activated olefins or from alkyl halides and the pentacyanocobaltate (II) anion[19,31,38,40]. The adduct obtained from butadiene is readily converted into a π-butenyl complex by expulsion of a cyanide ion[31]. Addition of cyanide ion converts the π-butenyl complex into the σ-butenyl one.

Reaction of the σ-butenyl complex with $[Co(CN)_5H]^{3-}$ gives mainly but-1-ene. The π-butenyl complex or an isomeric σ-butenyl complex in equilibrium with it yields mostly *trans*-but-2-ene. These observations explain the influence of cyanide ion concentration on the hydrogenation of butadiene.

The hydrogenation of phenyl vinyl ketone and phenacyl halides does not appear to follow this path as judged by the behaviour of the appropriate intermediates,

unless complicated by the formation of the π-oxaallyl complex which might be less reactive towards $[Co(CN)_5H]^{3-}$. In this context the postulation of a two-step hydrogen atom transfer mechanism, with the intermediate formation of free radicals, for the reduction of cinnamic acid may be noted[66]. Certainly complexes of the type $[Co(CN)_5R]^{3-}$ are not intermediates in the hydrogenation of alkyl and benzyl halides as even acid does not liberate the appropriate hydrocarbons[31], the corresponding nitriles being obtained instead[32]. Evidence for a free radical mechanism in these cases at least is compelling[31]. In particular, reduction of alkyl halides in the presence of acrylonitrile generates alkyl-acrylonitrile adducts and dimeric products are obtained from benzhydryl, trityl or tropylium halides. A kinetic study also supports a free-radical mechanism[20].

Information so far available indicates that the hexacyanodinickelate (I) anion can also catalyse homogeneous hydrogenation. The catalytic reduction of acetylene to ethylene has been patented[69] and an unstable complex with the apparent composition $K_4[Ni_2^{II}(CN)_6 C_2H_2]$ isolated[17]. The formation of a butadiene complex in solution has been observed and its decomposition found to liberate but-1-ene, *trans*-but-2-ene and *cis*-but-2-ene in the ratio 1:3.3:2.3[2]. The hydrogen here is derived from water possibly by way of an intermediate hydride. Similar species may well be involved in the promotion of borohydride reductions by tetracyanonickelate (II) ions[14], which generate hexacyanodinickelate (I) ions on reduction.

The catalytic activity of cobalt hydrotetracarbonyl in promoting homogeneous hydrogenation has already been discussed in Chapter 6. Apart from the reduction of aldehydes and the hydrogenolysis of benzyl alcohols already discussed a wide range of other systems can be hydrogenated under similar conditions, Table 10.2. Amongst the functional groups reduced are ketonic carbonyl, nitro and imino groups. The reduction of oximes, semicarbazides, nitriles and azo compounds has been encountered in Chapter VIII.

Attempts to hydroformylate aromatic compounds with "reactive double bonds" led to the discovery that while phenanthrene is only slightly reduced to 9,10-dihydrophenanthrene at 200°C, anthracene is quantitatively reduced to 9,10-dihydroanthracene at 135°C[12]. Amplification of this work has shown that a number of polycyclic hydrocarbons can be partially hydrogenated under these conditions; isolated benzene rings and phenanthrenoid systems being exceptionally resistant to reduction. The thiophene ring system is also reduced[15] under these conditions. In contrast to heterogeneous catalysts this catalytic system is much less susceptible to poisoning by sulphur containing compounds[33]. Indoles[62] and pyridines[59] are also sometimes reduced.

The use of other metal carbonyls as hydrogenation catalysts has been less extensively investigated. Iron carbonyls function as catalysts at temperatures around 200°C and hydrogen pressures of 200 to 350 atmospheres. Hydrogenation of the methyl esters of unsaturated fatty acids yields the saturated and monounsaturated esters[10,11,54]. The iron tricarbonyl complex derived from isomerised methyl linoleate has approximately the same catalytic activity[11,55,56]. Both pyridine and free fatty acid retard the hydrogenation[54,55]. Comparison with dicobalt octacarbonyl showed that the latter effects a faster reduction of conjugated and unconjugated double bonds, but that reduction stops at the monounsaturated ester stage[57]. The high activation energy of 60 kcal suggests that the rate determining step is thermal dissociation of the iron carbonyl[54]. A similar activation energy is found with the iron tricarbonyl complex[55,56]. It seems likely that a reaction sequence, analogous to that formulated for cobalt hydrotetracarbonyl catalysed hydrogenation, occurs.

TABLE 10.2 Hydrogenations Catalysed by Cobalt Hydrotetracarbonyl

Substrate	Product	Yield,(%)	References
Stilbene	Dibenzyl	—	16
Diphenylacetylene	Dibenzyl	80	16
Naphthalene	Tetralin	16	12
2-Methylnaphthalene	Methyltetralins	43	12
Acenaphthene	2a, 3, 4, 5-Tetrahydroacenaphthene	45	12
Fluorene	No reaction	—	12
Anthracene	9, 10-Dihydroanthracene	99	12
Phenanthrene	Mostly unchanged, small amounts di and tetrahydrophenanthrenes	—	12
Fluoranthene	1, 2, 3, 10b-Tetrahydrofluoranthene	54	12
Pyrene	4, 5-Dihydropyrene	69	12
Triphenylene	No reaction	—	12
Naphthacene	5, 12-Dihydronaphthacene	70	12
Chrysene	5, 6-Dihydrochrysene	24	12
1, 1-Dinaphthyl	No reaction	—	12
Perylene	1, 2, 3, 10, 11, 12-Hexahydroperylene	72	12

(Continued overleaf)

TABLE 10. 2 (continued)

Substrate	Product	Yield (%)	References
Coronene	No reaction	—	12
Benzyl alcohol	Toluene	63	6, 88
1-Phenylethanol	Ethylbenzene	70	6, 88
1-Naphthalenemethanol	1-Methylnaphthalene	72	88
Benzhydrol	Diphenylmethane	95	88
Triphenylcarbinol	Triphenylmethane	94	88
Benzopinacol	Diphenylmethane	89	88
Propionaldehyde	n-Propanol	57	89
n-Heptylaldehyde	n-Heptanol	50	6
Cyclohexylaldehyde	Cyclohexylcarbinol (46%) and formate (20%)	—	6
Benzaldehyde	Benzyl alcohol (24%), dibenzyl ether (54%)	—	6, 89
2-Phenylpropionaldehyde	2-Phenylpropan-1-ol	38	88
Diphenylacetaldehyde	2, 2-Diphenylethanol	88	88
Acrolein	Propionaldehyde	40-50	1
Crotonaldehyde	Butyraldehyde	40-50	1
	Butanol	—	89
Diethylketone	Limited reduction	—	6

Cyclohexanone	Limited reduction	—	6
Benzyl methyl ketone	Limited reduction	—	6
Acetophenone	Ethylbenzene	67	6, 88
p-Methoxyacetophenone	p-Ethylanisole	91	88
2-Acetylnaphthalene	2-Ethylnaphthalene	—	6
Benzophenone	Diphenylmethane	86	6, 88
Fluorenone	Fluorene	95	88
Benzanthrone	1,10-Trimethylenephenanthrene	59	88
Methyl vinyl ketone	Methyl ethyl ketone	70–90	1
Mesityl oxide	Methyl isobutyl ketone	70–90	1
Ethyl cinnamate	Ethyl β-phenylpropionate	70–90	1
Ethyl β-(2-furanyl) acrylate	Ethyl β-(2-furanyl) propionate	70–90	1
Thiophene	Thiolane	66	15
2-Methylthiophene	2-Methylthiolane	77	15
2-Ethylthiophene	2-Ethylthiolane	82	15
2,5-Dimethylthiophene	2,5-Dimethylthiolane	22	15
2-Thenyl alcohol	2-Methylthiophene (24%), 2-methylthiolane (57%)	—	15, 88
2-Acetylthiophene	2-Ethylthiophene (52%), 2-ethylthiolane (26%)	—	15

(Continued overleaf)

TABLE 10.2 (continued)

Substrate	Product	Yield (%)	References
Nitrobenzene	Diphenylurea	5	43
Acrylonitrile	2-Methylglutaronitrile	—	61
N-Cyclohexylidenecyclohexylamine	Dicyclohexylamine	37	53
N-Hexahydrobenzylidenecyclo-hexylamine	N-Hexahydrobenzylcyclohexylamine	30	53
N-Cyclohexylideneaniline	N-Cyclohexylaniline	59	53
N-Benzylidenecyclohexylamine	N-Benzylcyclohexylamine	60	53
N-Benzylideneaniline	N-Benzylaniline	80	6, 21, 43, 44, 53
N-Benzylidene-p-toluidine	N-Benzyl-p-toluidine	82	42, 43, 44
N-Benzylidene-2, 6-diethylaniline	N-Benzyl-2, 6-diethylaniline	82	21
N-Benzylidene-p-chloroaniline	N-Benzyl-p-chloroaniline	81	42, 43, 44
N-Benzylidene-p-anisidine	N-Benzyl-p-anisidine	83	42, 43, 44
N-Benzylidene-p-nitroaniline	N-Benzyl-p-nitroaniline	80	42, 43, 44
2-Methylindole	2, 3-Dihydro-2-methylindole	26	62
2-Phenylindole	2, 3-Dihydro-2-phenylindole	9	62
Pyridine	N-Formyl- and N-methyl-piperidine	—	59

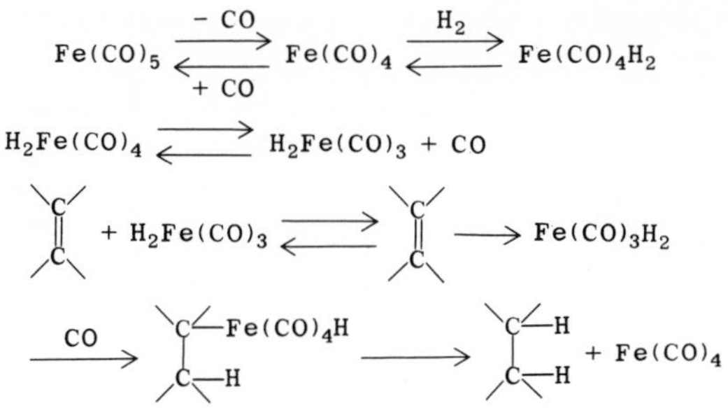

$$Fe(CO)_5 \underset{+\,CO}{\overset{-\,CO}{\rightleftarrows}} Fe(CO)_4 \underset{}{\overset{H_2}{\rightleftarrows}} Fe(CO)_4H_2$$

$$H_2Fe(CO)_4 \rightleftarrows H_2Fe(CO)_3 + CO$$

Iron carbonyls catalyse the reduction of nitriles to a mixture of primary and secondary amines[35]. The secondary amine can be envisaged as arising as formally depicted below.

$$R C \equiv N \longrightarrow RCH = NH \longrightarrow RCH_2NH_2$$
$$\downarrow R.\,CH_2NH_2$$
$$RCH = NCH_2R \longrightarrow RCH_2NH.\,CH_2R$$

Nickel carbonyl also catalyses this reduction.

A potentially useful catalytic specificity is shown by bis-(π-cyclopentadienyl) titanium dicarbonyl[68]. In general it seems to be a fairly selective catalyst for the hydrogenation of terminal acetylenes, Table 10.3. The hydrogenations proceed in hydrocarbon solvents at 50 to 60°C and 50 atmospheres. Conjugated dienes and dialkyl acetylenes are not reduced. The relative activating effects of substituents in terminal acetylenes is phenyl $> t.$butyl $>$ alkyl.

Ziegler-type systems also function as homogeneous hydrogenation catalysts[34,63,67]. Catalysts derived from cobalt (II and III), chromium (III), copper (II), iron (III), manganese (II and III), molybdenum (VI), nickel (II), palladium (II), ruthenium (III), titanium (IV) and vanadium (V) have been examined usually as acetylacetonates or alkoxides since the halides give poorer catalysts[67]. All but copper (II) give effective catalytic systems. Bis-(π-cyclopentadienyl) dichloride derivatives of titanium and zirconium also yield satisfactory catalytic systems[64,67]. Of the wide range of alkylmetal derivatives examined the best are tri-isobutylaluminium, triethylaluminium and di-isobutylaluminium hydride[67], although another report[64] states that zinc and aluminium alkyls are unsatisfactory with bis-(π-cyclopentadienyl) titanium dichloride. The most active catalysts were provided by cobalt (III), iron (III) and chromium (III) acetylacetonates in decreasing order of reactivity[67]. The lower reactivities found with cobalt (II), nickel (II) and palladium (II) may arise from the use of bis-phosphine complexes, as triphenylphos-

TABLE 10.3 Hydrogenations Catalysed by
$(\pi\text{-}C_5H_5)_2Ti(CO)_2$[68]

Substrate	Products
Styrene	Ethylbenzene
trans-Stilbene	Not reduced
Butadiene	Not reduced
Cycloocta-1, 3-diene	Not reduced
Acetylene	Not reduced
Pent-1-yne	Pent-1-ene (95%)
Hex-1-yne	Hex-1-ene (90%)
t.-Butylacetylene	*t*.-Butylethylene (40%) 2, 2-Dimethylbutane (60%)
Hept-3-yne	Not reduced
Phenylacetylene	Ethylbenzene (95%)
Diphenylacetylene	Dibenzyl (90%)

phine has been found to inhibit the activity of catalyst systems derived from nickel 2-ethylhexanoate and triethylaluminium[34].

Cyclohexene, hex-1-ene, oct-1-ene, 2-methylbut-2-ene, pent-2-ene, tetramethylethylene and stilbene are all reduced at temperatures of 25 to 40°C and a hydrogen pressure of 3·7 atmospheres[67]. The order of decreasing rate of hydrogenation of olefins is monosubstituted ~ disubstituted > trisubstituted > tetrasubstituted. In contrast to results with heterogeneous catalysts cyclohexene is hydrogenated faster than oct-1-ene. Phenylacetylene is reduced to ethylbenzene, but with hex-3-yne reduction is accompanied by cyclotrimerisation to hexaethylbenzene. Groups not reduced include aldehydes, ketones, esters, nitriles, nitro and azo. In common with other homogeneous hydrogenation catalysts metal hydride species are probably involved in these reductions. Direct one-step addition of molecular hydrogen is excluded by the observation that reduction of 2-methylbut-2-ene yields mono, di- and tri-deuterated products showing that hydrogen exchange occurs.

Aromatic nuclei are reduced under more vigorous conditions, 150-210°C and 70 atmospheres, using a catalyst derived from nickel (II) 2-ethylhexanoate and triethylaluminium, Table 10.4[34]. This

TABLE 10.4 Hydrogenation of Aromatic Compounds Using a
Ziegler-type Nickel Catalyst[34]

Substrate	Products	Yield, %
Benzene	Cyclohexane	100
o-Xylene	cis-1, 2-Dimethylcyclohexane	65·5
	trans-1, 2-Dimethylcyclohexane	34·5
Naphthalene	Tetralin	84
	Decalin	13
Phenol	Cyclohexanol	92
	Cyclohexanone	5
Dimethyl phthalate	Dimethyl hexahydrophthalate	100
Dimethyl terephthalate	Dimethyl hexahydroterephthalate	100
Pyridine	Piperidine	98

catalytic system is more reactive than Raney nickel, and in contrast
to normal experience fails to hydrogenate either nitrobenzene or
p-nitrophenol.

Comparison of analogous catalytic systems derived from other
transition metal carboxylates shows that the order of activity for
benzene hydrogenation is nickel ≥ cobalt > iron > chromium >
copper.

REFERENCES

1. Adkins, H. and Krsek, G., J. Amer. Chem. Soc., 1949, **71**, 3051.

2. Burnett, M. G., Chem. Comm., 1965, 507.

3. Calvin, M., Trans. Faraday Soc., 1938, **34**, 1181.

4. Calvin, M. and Wilmarth, W. K., J. Amer. Chem. Soc., 1956, **78**, 1301.

5. Cramer, R. D., Jenner, E. L., Lindsey, R. V. and Stolberg, U. G., J. Amer. Chem. Soc., 1963, **85**, 1691.

6. Dawydoff, V. W., Chem. Tech. (Berlin), 1959, **11**, 431.

7. Emken, E.A., Frankel, E. N. and Butterfield, R. O., J. Amer. Oil Chemists' Soc., 1966, **43**, 14.

8. Evans, D., Osborn, J. A., Jardine, F. H. and Wilkinson, G., Nature, 1965, **208**, 1203.

9. Flynn, J. H. and Hulburt, H. M., J. Amer. Chem. Soc., 1954, **76**, 3393.

10. Frankel, E. N., Emken, E. A. and Davison, V. L., J. Org. Chem., 1965, **30**, 2739.

11. Frankel, E. N., Emken, E. A., Peters, H. M., Davison, V. L., and Butterfield, R. O., J. Org. Chem., 1964, **29**, 3292.

12. Friedman, S., Metlin, S., Svedi, A. and Wender, I., J. Org. Chem., 1959, **24**, 1287.

13. Gillard, R. D., Osborn, J. A., Stockwell, P. B. and Wilkinson, G., Proc. Chem. Soc., 1964, 284.

14. Goerrig, D. and Weise, E., G. P. 1, 099, 506; Chem. Abs., 1962, **56**, 13806.

15. Greenfield, H., Metlin, S., Orchin, M. and Wender, I., J. Org. Chem., 1958, **23**. 1054.

16. Greenfield, H., Wotiz, J. H. and Wender, I., J. Org. Chem., 1957, **22**, 542.

17. Griffith, W. P. and Wilkinson, G., J. Chem. Soc., 1959, 1629.

18. Halpern, J., Harrod, J. F. and James, B. R., J. Amer. Chem. Soc., 1961, **83**, 753; Ann. Rev. Phys. Chem., 1965, **16**, 108.

19. Halpern, J. and Maher, J. P., J. Amer. Chem. Soc., 1964, **86**, 2311.

20. Halpern, J. and Maher, J. P., J. Amer. Chem. Soc., 1965, **87**, 5361.

21. Horiie, S. and Murahashi, S., Bull. Chem. Soc. Japan, 1960, **33**, 247.

22. Isogai, K. and Hazeyama, Y., J. Chem. Soc. Japan, 1965, **86**, 869.

23. James, B. R. and Rempel, G. L., Canad. J. Chem., 1966, **44**, 233.

24. Kang, J., J. Chem. Soc. Japan, 1963, **84**, 56.

25. King, N. K. and Winfield, M. E., J. Amer. Chem. Soc., 1961, **83** 3366.

26. Kwiatek, J. and Mador, I. L., U.S.P. 3, 185, 727; Chem. Abs., 1965, **63**, 2900.

27. Kwiatek, J. and Mador, I. L., U.S.P. 3, 205, 217; Chem. Abs., 1965, **63**, 17978.

28. Kwiatek, J. and Mador, I. L., U.S.P. 3, 205, 266; Chem. Abs., 1965, **63**, 17979.

29. Kwiatek, J., Mador, I. L. and Seyler, J. K., J. Amer. Chem. Soc., 1962, **84**, 304.

30. Kwiatek, J., Mador, I. L. and Seyler, J. K., Adv. Chem. Series, 1963, **37**, 201.

31. Kwiatek, J. and Seyler, J. K., J. Organometallic Chem., 1965, **3**, 421.

32. Kwiatek, J. and Seyler, J. K., J. Organometallic Chem., 1965, **3**, 433.

33. Laky, J., Szabo, P. and Marko, L., Acta Chim. Acad. Sci. Hung., 1965, **46**, 247.

34. Lapporte, S. J. and Schuett, W. R., J. Org. Chem., 1963, **28**, 1947.

35. Levering, D. R., U.S.P. 3, 152, 184; Chem. Abs., 1965, **62**, 427.

36. Mabrouk, A. F., Dutton, H. J. and Cowan, J. C., J. Amer. Oil Chemists' Soc., 1964, **41**, 153.

37. Mabrouk, A. F., Selke, E., Rohwedder, W. K. and Dutton, H. J., J. Amer. Oil Chemists' Soc., 1965, **42**, 432.

38. Mason, R. and Russell, D. R., Chem. Comm., 1965, 182.

39. Maxted, E. B. and Ismail, S. M., J. Chem. Soc., 1964, 1750.

40. Mays, M. J. and Wilkinson, G., Nature, 1964, **203**, 1167.

41. Mizuta, T. and Kwan, T., J. Chem. Soc. Japan, 1965, **86**, 1010.

42. Murahashi, S. and Horiie, S., Ann. Report Sci. Works, Fac. Sci., Osaka Univ., 1959, **7**, 89.

43. Murahashi, S. and Horiie, S., Bull. Chem. Soc. Japan, 1960, **33**, 78.

44. Murahashi, Horiie, S. and Jo, T., J. Chem. Soc. Japan, 1958, **79**, 68.

45. Murakami, M. and Kang, J., Bull. Chem. Soc. Japan, 1962, **35**, 1243

46. Murakami, M. and Kang, J., Bull. Chem. Soc. Japan, 1963, **36**, 763.

47. Murakami, M., Kang, J., Itatani, H., Senoh, S. and Matsusato, N., J. Chem. Soc. Japan, 1963, **84**, 48.

48. Murakami, M., Kang, J., Itatani, H., Senoh, S. and Matsusato, N., J. Chem. Soc. Japan, 1963, **84**, 51.

49. Murakami, M., Kang, J., Itatani, H., Senoh, S. and Matsusato, N., J. Chem. Soc. Japan, 1963, **84**, 53.

50. Murakami, M., Kawai, R. and Suzuki, K., J. Chem. Soc. Japan, 1963, **84**, 669.

51. Murakami, M., Suzuki, K., Itatani, H., Kyo, M. and Senoh, S., Jap. P. 21, 974('65); Chem. Abs., 1966, **64**, 4945.

52. Murakami, M., Suzuki, K. and Kang, J., J. Chem. Soc. Japan, 1962, **83**, 1226.

53. Nakamura, A. and Hagihara, N., Mem. Inst. Sci. Ind. Research, Osaka Univ., 1958, **15**, 195.

54. Ogata, I. and Misono, A., Yukagaku, 1964, **13**, 644; Chem. Abs., 1965, **63**, 17828.

55. Ogata, I. and Misono, A., J. Chem. Soc. Japan, 1964, **85**, 748.

56. Ogata, I. and Misono, A., J. Chem. Soc. Japan, 1964, **85**, 753.

57. Ogata, I. and Misono, A., Yukagaku, 1965, **14**, 16; Chem. Abs., 1965, **63**, 17828.

58. Osborn, J. A., Wilkinson, G. and Young, J. F., Chem. Comm., 1965, 17.

59. Pino, P. and Ercoli, R., Ricerca Sci., 1953, **23**, 1231.

60. Rylander, P. N., Himelstein, N., Steel, D. R. and Kreidl, J., Engelhard Ind. Tech. Bull., 1962, **3**, 61; Chem. Abs., 1962, **57**, 15864.

61. Schreyer, R. C., U.S.P. 3, 206, 498; Chem. Abs., 1966, **64**, 605.

62. Shaw, J. T. and Ryson, F. T., J. Amer. Chem. Soc., 1956, **78**, 2538.

63. Shell Internationale Research Maatschappij N.V., Neth. Appl. 296, 137; Chem. Abs., 1965, **63**, 9878.

64. Shikata, K., Nishino, K., Azuma, K. and Takegami, Y., J. Chem. Soc. Japan, Ind. Chem. Sect., 1965, **68**, 358.

65. Simandi, L. and Nagy, F., Acta Chim. Acad. Sci. Hung., 1965. **46**, 101.

66. Simandi, L. and Nagy, F., Acta Chim. Acad. Sci. Hung., 1965, **46**, 137.

67. Sloan, M. F., Matlock, A. S. and Brelow, D. S., J. Amer. Chem. Soc., 1963, **85**, 4014.

68. Sonogashira, K. and Hagihara, N., Bull. Chem. Soc. Japan, 1966, **39**, 1178.

69. Spencer, M. S., U.S.P. 2, 966, 534; Chem. Abs., 1961, **55**, 8288.

70. Stouthamer, B. and Flugter, J. C., J. Amer. Oil Chemists' Soc., 1965, **42**, 646.

71. Suzuki, T. and Kwan, T., J. Chem. Soc. Japan, 1965, **86**, 713.

72. Suzuki, T. and Kwan, T., J. Chem. Soc. Japan, 1965, **86**, 1198.

73. Suzuki, T. and Kwan, T., J. Chem. Soc. Japan, 1965, 86, 1341.

74. Suzuki, T. and Kwan, T., J. Chem. Soc. Japan, 1966, **87**, 395.

75. Tulupov, V. A., Zhur. fiz. Khim., 1957, **31**, 519.

76. Tulupov, V. A., Zhur. fiz. Khim., 1958, **32**, 727.

77. Tulupov, V. A., Zhur. fiz. Khim., 1962, **36**, 1617.

78. Tulupov, V. A., Zhur. fiz. Khim., 1963, **37**, 698.

79. Tulupov, V. A., Zhur. fiz. Khim., 1964, **38**, 1059.

80. Tulupov, V. A. and Evlesheva, T. I., Zhur. fiz. Khim., 1965, **39**, 84.

81. Tulupov, V. A. and Gagarina, M. I., Zhur. fiz. Khim., 1964, **38**, 1695.

82. Tulupova, A. I. and Tulupov, V. A., Zhur. fiz. Khim., 1963, **37**, 2678.

83. Vaska, L., Inorg. Nuclear Chem. Letters, 1965, **1**, 89.

84. Vaska, L. and Rhodes, R. E., J. Amer. Chem. Soc., 1965, **87**, 4970.

85. Vries, B. de, Proc., k. ned. Akad. Wetenschap., 1960, **63B**, 443.

86. Vries, B. de, J. Catalysis, 1962, **1**, 489.

87. Weller, S. and Mills, G. A., J. Amer. Chem. Soc., 1953, **75**, 769.

88. Wender, I., Greenfield, H. and Orchin, M., J. Amer. Chem. Soc., 1951, **73**, 2656.

89. Wender, I. and Orchin, M., U.S.P. 2, 614, 107; Chem. Abs., 1953, **47**, 5422.

90. Wright, L. W. and Weller, S., J. Amer. Chem. Soc., 1954, **76**, 3345.

91. Young, J. F., Osborn, J. A., Jardine, F. H. and Wilkinson, G., Chem. Comm., 1965, 131.

INDEX